NATURA

Biology for Bilingual Classes

edited by

Doris Bächle-Knauer
Susanna Bächle

Genetics and Immune System

Ernst Klett Verlag
Stuttgart · Leipzig

Warning signs and experiments in school

Natural sciences like biology are not imaginable without experiments. Natura Oberstufe contains also some experiments.

However, experimenting with chemicals is never completely safe. Therefore, it is important to discuss potential sources of danger with your teacher before starting the respective experiment. Especially in the laboratory obvious behavioural rules always have to be observed. The precautions taken are dependent on the danger potential of the used substance.

For that reason the chemicals are marked with a warning sign in the experimental manual. The warning signs are also present on the label of the chemical's container.

They mean:

C = *corrosive*
Living tissue and materials that come in contact with this substance will be destroyed at the affected site.

F = *flammable*
Substances that are easily ignited when exposed to an ignition source.

Xi = *irritating* (X stands for St. Andrew's cross)
Substances that can be irritating to skin, eyes or respiratory organs.

Xn = *noxious*
Substances that can cause health problems when inhaled, swallowed or contacted the skin.

1. Auflage   1 $^{9\ 8\ 7\ 6}$ | 22 21 20

Alle Drucke dieser Auflage sind unverändert und können im Unterricht nebeneinander verwendet werden.
Die letzte Zahl bezeichnet das Jahr des Druckes.

Bearbeitet von: Doris Bächle-Knauer; Max-Planck-Gymnasium, Schorndorf; Staatl. Seminar für Didaktik und Lehrerbildung, Stuttgart; Susanna Bächle; B. Sc. Molekulare Medizin, Karolinska Institutet, Stockholm
unter Mitarbeit von: Dr. Irmtraud Beyer, Dreieich; Dr. Horst Bickel, Düsseldorf; Roman Claus, Rees; Roland Frank, Stuttgart; Prof. Dr. Harald Gropengießer, Hannover; Gert Haala, Wesel, Prof. Dr. Siegfried Kluge; Neumark; Bernhard Knauer, Göttingen; Dr. Inge Kronberg, Hohenwestedt; Hans-Peter Krull, Kaarst; Hans-Dieter Lichtner, Stadthagen; Uschi Loth, Burbach; Dr. Horst Schneeweiß, Hamburg; Dr. Jürgen Schweizer, Stuttgart; Ulrich Sommermann, Münchberg; Gerhard Ströhla, Münchberg; Dr. Wolfgang Tischer, Sarstedt; Günther Wichert, Dinslaken

Redaktion: Dr. Peter R. Menke
Mediengestaltung: Ingrid Walter

Fachwissenschaftliche und sprachliche Beraterin: Dr. R. Theresa Jones, Wolfenbüttel
Lautschrift: Peter Bereza, Aachen
Gestaltung: Prof. Jürgen Wirth, Visuelle Kommunikation, Dreieich
Umschlaggestaltung: normaldesign GbR, Schwäbisch Gmünd
Illustrationen: Prof. Jürgen Wirth, Visuelle Kommunikation, Dreieich; unter Mitarbeit von Matthias Balonier, Evelyn Junqueira, Nora Wirth
Reproduktion: Meyle + Müller, Medien-Management, Pforzheim
Druck: PASSAVIA Druckservice GmbH & Co. KG, Passau

Printed in Germany
ISBN 978-3-12-045321-5

In this book you will find all of the topics taught in genetics lessons in the "Oberstufe". Biology is a subject that looks at the multiple variants of life in nature. For this reason, the topics and examples in this book are very varied.

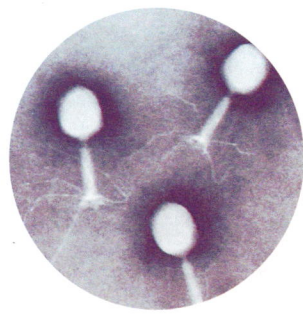

In order to make this process easier, this book contains many pages which not only provide information, but also motivate students to discuss, explore and question scientific topics.

Info pages
These provide basic information about a topic. Numerous figures illustrate complex issues. The short tasks test whether you have understood the text, or not.

CD
The answers to the tasks are on a CD, which also contains word lists (with pronunciation) to help you to learn in a foreign language.

»info box«
In info boxes you find interesting examples, exceptions, methods, etc.

Practicals pages
Here experiments are described that you can carry out yourself. It is important to work precisely and to make accurate reports in order to obtain the best results.

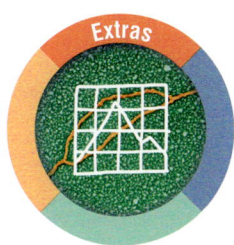

Extras pages
Extensive information about a specific topic and corresponding tasks are provided on these pages.

Encyclopedia pages
These pages provide additional information which place the topics you have learned about in a broader context and in that way give an overview. Topics are included that are not mentioned directly in the teaching curriculum. These pages are meant to encourage further reading.

Issues pages
Biological problems often go beyond the subject of biology.
These pages provide interdisciplinary material and impulses to work independently. They also refer to topics in everyday life.

Basic concepts
Biology is a complex science and its different disciplines are connected by many, partly abstract, relations. This new type of page shows interconnections between very different topics. Basic concept pages clarify principles and enable you to order and structure known facts. The tasks encourage pupils to discuss and to think about examples again.

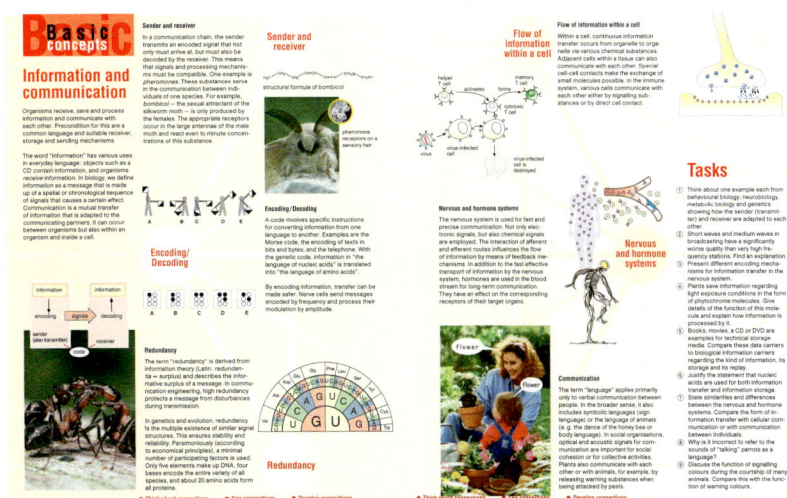

Foreword

Our knowledge of the natural sciences is constantly developing. Nowadays, the focus has shifted more to the molecular level of processes, with many explanations for the "large picture" having been found by the examination of small molecules and their interactions. However, the observation and description of the visible world is still a major part of research. The aim of this book is to give an overview of the "large picture" of selected biological principles focusing mainly on DNA and its functions, on reproduction, on genetics and on embryonic development. Furthermore, the book introduces pupils to the quickly advancing field of biotechnology. The last chapter is devoted to the cunning defence system of the human body — immunology. We have tried to explain the molecular basics of immunology and an attempt has been made to simplify complex issues without losing accuracy.

Genetics includes the scientific studies of heredity and hereditary variation. This biological discipline is based upon the fact that living organisms are distinguished by their ability to reproduce their own kind. Obviously, offspring resemble their parents more than they do less closely related individuals.

For thousands of years, farmers have exploited breeding plants and animals for desired features. This strategy represents the intuitive understanding of human beings and now also relies on "genetic know-how". Various topics will be presented in this book at three different levels: molecules, cells and organisms.

The Augustian monk, JOHANN GREGOR MENDEL, published his "experiments with plant hybrids" in 1866 — seven years after CHARLES DARWIN had published his "origin of species". From more than 10,000 crossing experiments, Mendel determined ratios from which he derived the laws of inheritance. Unfortunately, their *general* validity was not recognized by his contemporaries. Why did MENDEL work with peas? Because they are available in many varieties. A "character" is defined as a heritable feature, such as colour, that varies among individual organisms. However, each variant of a character, such as white or red, is called a "trait". It is useful to distinguish between these two terms.

In 1869, three years later, FRIEDRICH MIESCHER isolated "nuclein" — from nuclei — for the very first time at the castle of Tübingen. German-speaking scientists have contributed, over the last century, essential hypotheses for genetic research. Nowadays, their results would be published in English. One prominent example of such a German research worker is CHRISTIANE NÜSSLEIN-VOLHARD; she and her colleagues have clarified basic genetic mechanisms essential for embryonic development. Her papers, in English, have been published, for example, by the famous international weekly journal of science, "Nature". In 1995, NÜSSLEIN-VOLHARD'S achievements were awarded with the Nobel Prize.

Molecular genetics is revolutionizing agriculture and medicine. Geneticists have made important contributions to elucidating the way in which multicellular animals and plants arise from a single cell. All areas of the biosciences are influenced by genetics — from cell biology to immunology, ecology, evolutionary biology, and even behavioural biology. In this book, pupils will be also asked to consider ethical questions raised by our ability to manipulate nucleic acids.

Organisms must defend themselves against the many potentially dangerous viruses, bacteria and other pathogens that they encounter. Animals (and humans) must also contend with abnormal body cells that may develop into cancer. The topic "acquired immunodeficiency syndrome" (= AIDS), for example, is of vital importance for *global* health. The basic functions of the immune system will therefore be explained and what happens when this system goes wrong.

After working with this book, pupils should be able to understand the basic principles of DNA and genetics. In addition, they should be aware of the main ideas of biotechnology and immunology.

Bilingual classes are a challenge for both pupils and teachers. As mentioned above, most scientific articles are written in English, with the language providing a link between researchers around the world and the knowledge that they have. One can say that English has become the "lingua franca" of science. Therefore, it is essential to be comfortable with English and to learn the commonly used scientific terms. This book aims at taking away the scepticism, and maybe even the fear, that teaching and learning in a foreign language might cause. The level of English used in the book corresponds to that expected in the "Oberstufe". Additionally, pupils should find the word lists for each topic and the accompanying worksheets (on the CD) helpful. The layout of this book corresponds to chapter 3 of the German textbook, "NATURA, Biologie für Gymnasien, Oberstufe", which can always be used for reference purposes.

The editors of the book hope that they have succeeded in providing scientific knowledge while helping pupils to learning in English. We wish both the pupils and teachers fun with this book and the fascinating subject of biology.

Doris Bächle-Knauer
Susanna Bächle

October 2008

Contents

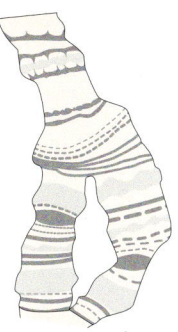

Genetics

Gene therapy, genetic engineering, gene sequencing, gene map, the genome project, genetic fingerprinting — genetics and its applications are present in discussion rounds and the daily press. Despite not even being 150 years old, genetics is a comparably young science. GREGOR MENDEL is known as the founder of genetics. His cross-breeding experiments with pea plants revealed the rules of inheritance. The word "gene" was not used by him; he called the inheritable trait simply a "factor".

The fast development of genetics is mirrored surprisingly well in the change of the definition of the term "gene": The classical approach acts on the assumption of visible traits and understands genes as undividable units of inheritance. The discovery of the structure and function of DNA has made genes approachable at a molecular level: genes are defined segments of DNA responsible for the expression of traits. This makes it possible to follow traits of MENDEL's peas right down to the DNA in the nucleus.

Meanwhile, the complete DNA sequences of many organisms have been mapped. However, it is still not possible to determine how many genes are included in a genome. The reason for this is the presence of overlapping, repetitive and non-informative sequences. So, on the one hand, high-tech methods help us to isolate substances and analyse genes much better but, on the other hand, our understanding of gene function has become more and more abstract.

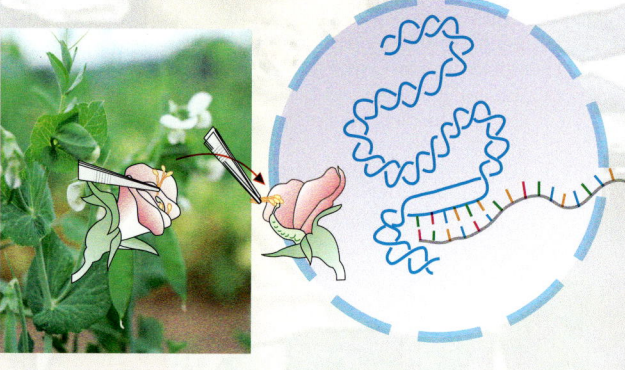

"Soweit die Erfahrung reicht, finden wir es überall bestätigt, dass constante Nachkommen nur dann gebildet werden können, wenn die Keimzellen und der befruchtende Pollen gleichartig, somit beide mit der Anlage ausgerüstet sind, völlig gleiche Individuen zu beleben, wie das bei der normalen Befruchtung der reinen Arten der Fall ist."
GREGOR MENDEL 1866

"A gene is the segment of DNA which codes for an m-RNA molecule."
FRANÇOIS JACOB and JACQUES MONOD 1961

F_1

"A gene is a combination of DNA segments that together constitute an expressible unit, expression leading to the formation of one or more functional gene products that may be either RNA molecules or polypeptides. The segments of a gene include the transcribed region, which encompasses the coding sequences, intervening sequences, any 5'leader and 3'trailer sequences that surround the ends of the coding sequences, and any regulatory segments included in the transcription unit, and the regulatory sequences that flank the transcription unit and are required for specific expression."
MAXINE SINGER and PAUL BERG 1991

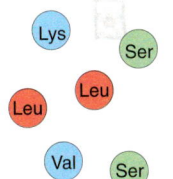

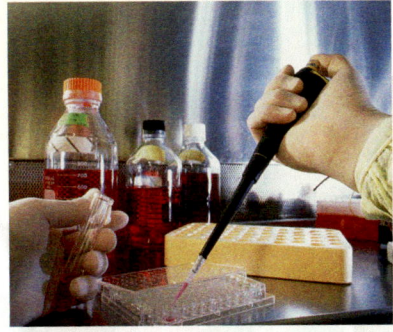

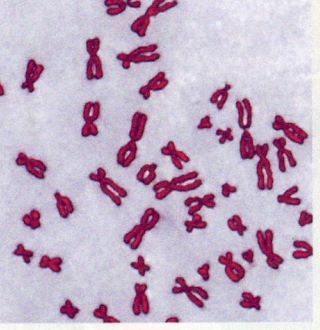

"The word 'gene' is completely free from any hypotheses; it expresses only the evident fact that, in any case, many characteristics of the organism are specified in the gametes by means of special conditions, foundations, and determiners which are present in unique separate, and thereby independent ways — in short, precisely what we wish to call genes."
WILHLELM JOHANNSEN 1909

"Each gene controls the production, function and specifity of a particular enzyme."
GEORGE W. BEADLE & EDWARD TATUM 1941

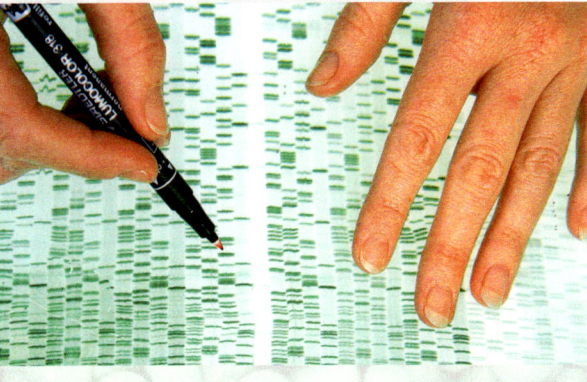

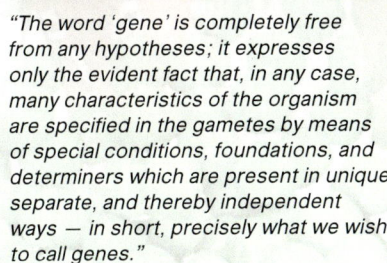

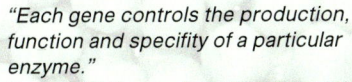

F₂

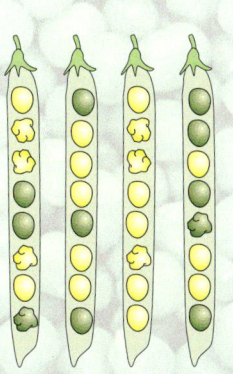

7

The carrier of hereditary information – experimental evidence

slime layer or
slimy capsule

cell membrane
(7.5 – 8 nm thick)

cell wall
(10 – 80 nm thick)

A typical bacterial cell with its
outer layers
(1 nm = 1 millionth mm)

DNA
Deoxyribo**n**ucleic **a**cid

Transformation
The transfer of DNA
into any living cell

In the 30s of the last century, the question as to whether proteins or DNA carry the genetic information could not be answered. Both substances consist of long linear chain-like molecules that can form an almost infinite number of variations.

The first experiments analysing the chemical nature of the carrier of genetic information were performed on bacteria. In 1928, FREDERICK GRIFFITH used the bacterium *Streptococcus pneumoniae*, which exists in two variants. The S strain is able to produce a slimy polysaccharide capsule. It forms smooth colonies (S for smooth). The S strain is virulent, which means that it causes deadly pneumonia in mice. If these bacteria are killed by heat, they lose their virulence. Because of genetic differences, however, the R strain cannot produce a slimy capsule. Within a bacteria culture it forms rough colonies (R for rough) and is not virulent.

GRIFFITH mixed dead S strain bacteria with living R strain bacteria. Surprisingly, this mixture could cause pneumonia in mice. It was even possible to isolate living S strain bacteria from the ill mice. The ability to produce slimy capsules had been transferred from the dead S bacteria to the living R bacteria. The genetic information is therefore based onto a transferable substance.

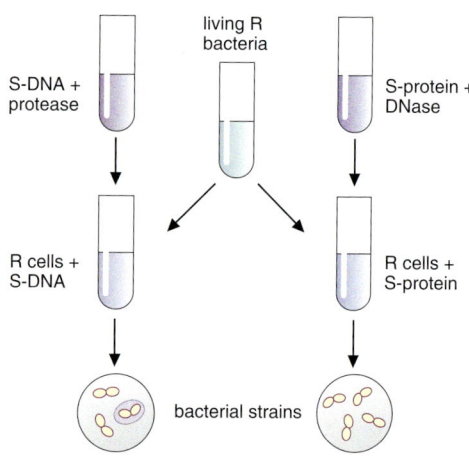

2 AVERY: DNA is the transforming substance

In 1944, the bacteriologist OSWALD AVERY resumed GRIFFITH's experiments to identify the substance that was transferred from S to R cells. He separated DNA and proteins from deadened S cells and mixed living R cells either with the S strain DNA or S strain proteins. The S strain proteins did not enable the R cells to produce slimy capsules. The S strain DNA, however, transferred the ability to produce a capsule to the R strain bacteria. With these experiments, AVERY had established that the DNA is the substance carrying genetic information. The transfer of DNA into living cells is called *transformation*.

OSWALD AVERY
(1877 – 1955)

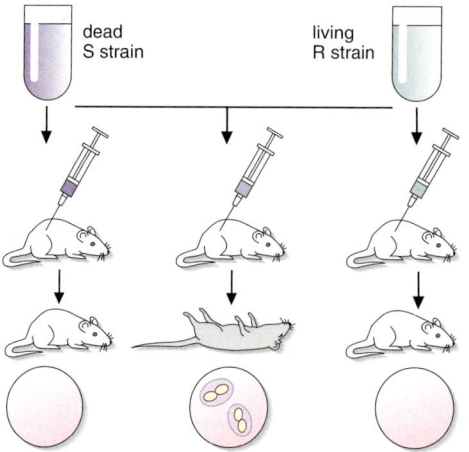

1 GRIFFITH's experiments

Tasks

① OSWALD AVERY performed several control experiments. For example, he treated the separated DNA with protein-cleaving enzymes (*proteases*) and the separated proteins with DNA-cleaving enzymes (*DNases*, deoxyribonucleases). Explain why.

② AVERY added serum to the bacterial cultures. Serum treated with heat did not interfere with his experiments. Untreated serum, however, often inhibited the transformation of R cells by material isolated from S cells. Serum contains a number of different enzymes. Explain the phenomenon.

Experiments with DNA

Attaining DNA

In molecular biology laboratories, nucleic acid precipitation is a commonly used method. Often, the extraction of DNA from homogenised tissue is the basis for further genetic analyses. The inherited material DNA can be extracted easily from cells by means of simple chemical methods to isolate and separate the DNA from other cellular components.

Materials:
Ethanol (deep frozen), sodium chloride (NaCl, table salt), tomato, washing-up liquid, centrifuge, glass stirrer, washing powder with proteases, knife, mortar, beaker, test tube, coffee filter

Experiment:
Take the knife and cut the tomato into large pieces. Mix the pieces together with 40 ml water, 5 ml washing-up liquid, a few particles of washing powder and half a teaspoon of table salt and homogenise them with a pestle and mortar. Put the homogenate into a beaker and keep it at 60°C in a water bath for 15 min.

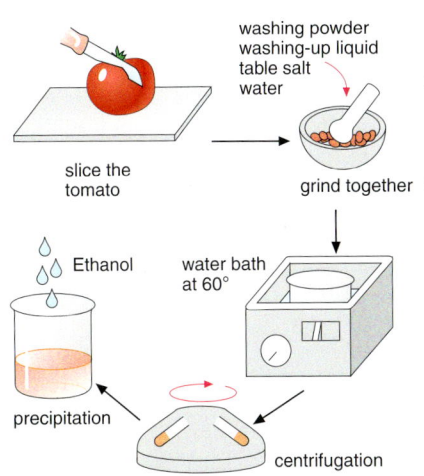

A sample of the cell-free extract is now transferred into a test tube and centrifuged to remove cell debris. The cell debris forms a sediment at the base of the tube. If the supernatant is not clear after centrifugation, it can be filtered additionally through a coffee filter.

To precipitate the DNA, carefully pour ethanol onto the supernatant so that it forms a separate layer on top. At the boundary at which the ethanol and water meet, a gel-like mass forms, the precipitated DNA. This can be wound around the glass stirrer.

Fragment production

The thread-like DNA molecules are often too long and bulky for further experimentation. Isolated DNA is therefore cleaved into smaller pieces (fragments). For this purpose, a number of DNA-cleaving enzymes are used nowadays. These *restriction enzymes* recognize a specific sequence of nucleotides within DNA strands and cut the DNA only at these points, the so-called restriction sites. Many short fragments of defined length are obtained. Unspecific DNases, however, have the ability to degrade the DNA completely into single *nucleotides*, beginning at the ends of the DNA strands.
Acidic hydrolysis can also degrade DNA chemically into nucleotides.

Experiment:
The DNA isolated from the tomatoes is mixed with 10 ml sulfuric acid (1 mol/l) for chemical fragmentation in a test tube [C] (wear safety glasses!). The test tube is lightly covered with aluminium foil and held in boiling water for 15 min. The DNA is thereby cleaved into nucleotides - it "dissolves".

Tasks

(1) How can you experimentally disprove the claim that the isolated DNA is not hydrolysed by the acid but only switched again to its soluble form?
(2) In the DNA precipitation experiment, washing powder containing proteases is used. *Proteases* are protein-cleaving enzymes. Which proteins are associated with the DNA molecule? Consult this book.

Separation of fragments

If DNA is cut by restriction enzymes, it is often necessary for further experiments to make the fragments visible. Therefore, the fragments are separated on a starch-gel (*electrophoresis*). The charged DNA fragments move within the electric field, smaller fragments moving faster than larger ones. Similar fragments are concentrated in bands. These bands can be made visible by using dyes.

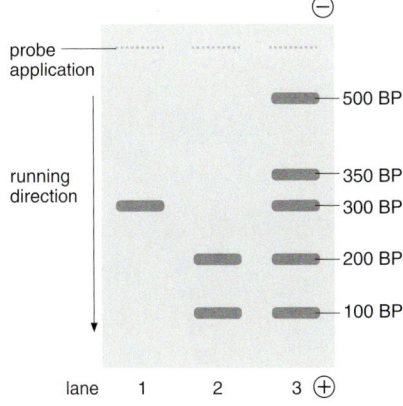

The above figure shows such a separation in a fragmentation experiment (lane 2). In order to estimate the size of the DNA fragments a so-called molecular weight standard (lane 3) is used. It contains DNA fragments of known size, indicated in bp (base pairs).

Tasks

(3) What is the charge of the separated DNA fragments? What size are they?
(4) How often has the employed restriction enzyme cut the sample DNA (lane 1) to give the DNA bands in lane 2?

Isolation of fragments: blotting

If DNA is cut by restriction enzymes, it is often necessary for further experiments to make the fragments visible. Therefore, the fragments are separated on a starch-gel (see page 21). The charged DNA fragments move within the electric field, smaller fragments moving faster than larger ones. Similar fragments are concentrated in bands. These bands can be made visible by using dyes.

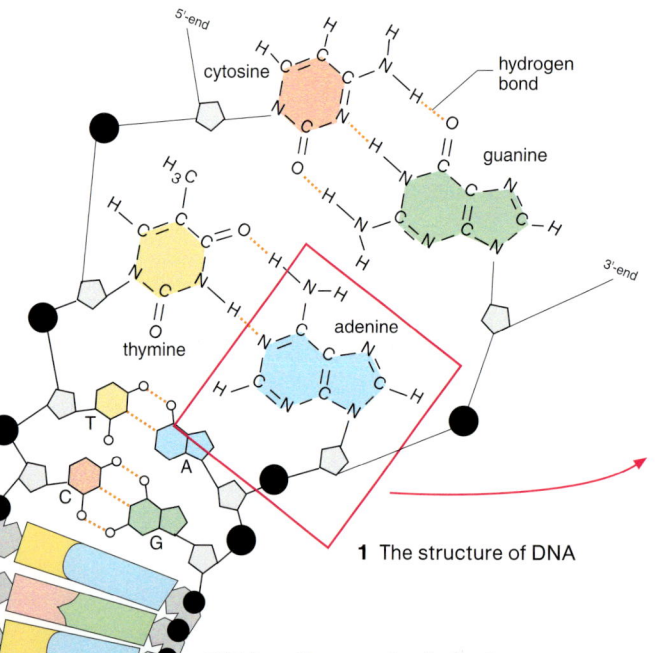

1 The structure of DNA

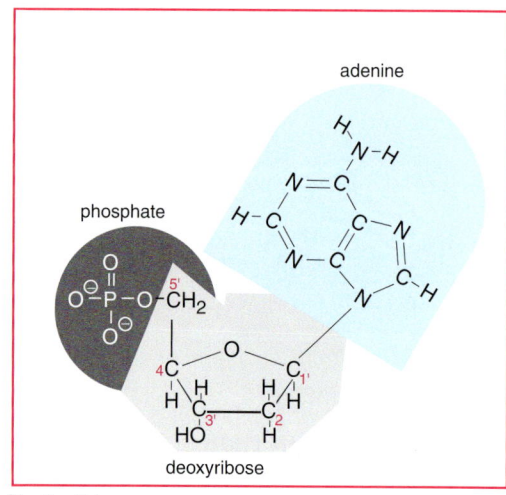

Nucleotid

DNA – the material of genes

Organisms are characterized by attributes such as flower colour, leaf shape, blood group, hair colour and many more. Chemicals in the nucleus control the expression of these attributes and are inherited by the next generation.

A molecule that carries and passes on genetic information needs the following characteristics:
— It has to encode information and possess a reading direction.
— It has to be duplicated unchanged and passed on to the next generation.
— It has to be stable but, at the same time, must allow slight modifications.

After decades of molecular-genetic detective work DNA (*deoxyribonucleic acid*) was identified as the carrier of the genetic information. The units of genetic information are the *genes*, which are segments on the DNA molecule.

Based on the experiments carried out by FRIEDRICH MIESCHER and ALBRECHT KOSSEL, it was known as early as the 19th century that nuclei contain nucleic acids made up in equal amounts of the C5-sugar deoxyribose, phosphate and the four bases *adenine (A), thymine (T), cytosine (C)* and *guanine (G)*.

A and G are purine bases, T and C are pyrimidine bases. When comparing the DNA of different organisms, ERWIN CHARGAFF (1951) found out that all DNA contained the same number of molecules of A and T and the same number of molecules of C and G (*Chargaff's rule*, fig.11.1).

X-ray examination of DNA crystals by ROSALIND FRANKLIN (1953) showed that DNA is composed of two into a spiral twisted single strands in which the phosphate molecules point outwards and the bases point inwards. JAMES D. WATSON and FRANCIS C. CRICK ingeniously solved the puzzle presented by these data and proposed the *DNA double helix model* in 1953. They realized that the two single strands are orientated in opposing directions and are intertwined. The two strands are held together by hydrogen bonds between the bases.

A few chemical details are needed to gain a better understanding of this structure: Every DNA strand is made up of a chain of DNA nucleotides. Each DNA nucleotide contains a sugar, phosphate and one of the four bases A, T, G or C. The C5 sugar deoxyribose can form 1', 3' and 5' bonds to other molecules through its carbon atoms.

ø 11 nm

The phosphate binds to the 3'C-atom of one sugar and to the 5'C-atom of the following sugar, thereby allowing a deoxyribose chain to be linked.

Each DNA strand therefore has a 5'-end for phosphate and a 3'-end with a free OH group. Thus, DNA has a reading direction. The two single strands are coiled around each other on a common axis to form a double helix. The 5'-end of one strand and the 3'-end of the other strand oppose each other (lie *antiparallel*). The bases bind to the C1 atom of the sugar and point inwards into the double strand where they bind to the bases of the other single strand. A (long) purine base always pairs with a (short) pyrimidine base. Adenine and thymine form two hydrogen bonds, whereas guanine and cytosine form three hydrogen bonds. These pairs are called *complementary base pairs* (A - T and G - C). This explains the base ratio of the Chargaff rule and the stability of double-stranded DNA.

The DNA molecule resembles a twisted rope-ladder whose rungs are made of base pairs and whose uprights are alternately made of phosphate and sugar molecules. The DNA nucleotide sequence (the sequence of bases in a single strand) encodes the genetic information. The *base sequence* can be simplified as a letter sequence (e.g. AAATT-CGAA). The DNA double helix model has led to a breakthrough in molecular genetics: today, the base sequences of many species have been identified.

DNA is packaged

The DNA of prokaryotes forms a closed ring. In eukaryotes, it is divided into separate linear units called *chromosomes*. The genetic information of a cell is very extensive: The overall DNA length of *Escherichia coli* is 1.5 mm (4.7 million base pairs) and, in each human cell, it is 97 cm (3 billion base pairs). It is amazing that such long molecules can be stored within a cell. In eukaryotes, DNA is compressed by a highly ordered packing procedure. Specific proteins called *histones* are involved in this process. Histone complexes function as pin curlers around which the DNA molecule is wrapped.

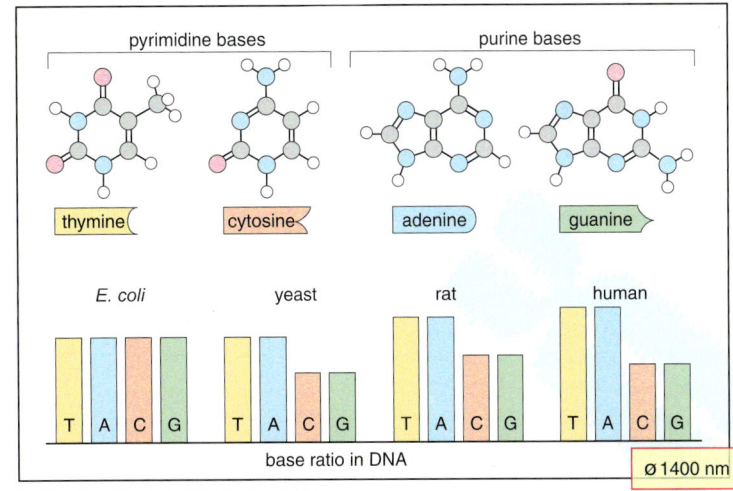

1 The four bases present in DNA

In this way, a *nucleosome* is generated. Nucleosomes appear in an electron microscope image as pearls on a string. The wrapping shortens the DNA by approximately $1/6$. The strand of nucleosomes is coiled again at regular intervals into a 30 nm thick fibre, which is then organized into loops and rosettes.

The DNA-histone complexes in the nucleus can be stained and are visible in the light microscope as *chromatin*. Highly condensed parts are stained more intensively and are called *heterochromatin*. They are considered to have no genetic activity. In the less packed and less stained *euchromatin*, the DNA is less condensed and genetically active.

In addition to the spatial difference, a temporal change in density also occurs. Just before each cell replication, the DNA must not only be duplicated, but also densely packed so that it can be distributed to the daughter cells without losses. For this purpose, the DNA rosettes become even more tightly coiled.

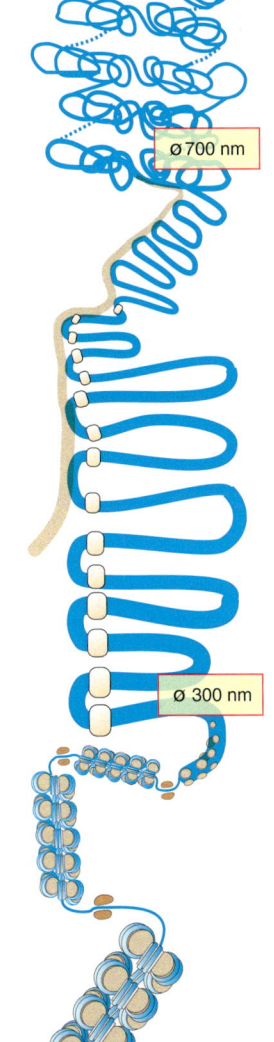

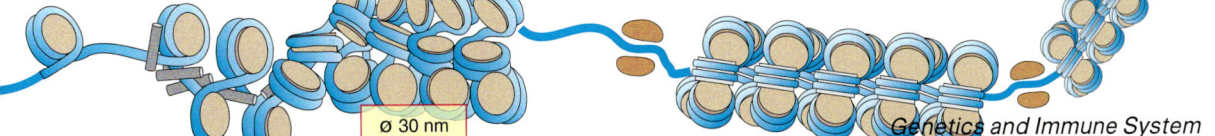

DNA-replication – from one make two

The DNA double helix fulfils all the requirements needed for carrying genetic information: It can encode this information through the base sequence, it has a reading direction and it is stable. WATSON and CRICK realized that also a duplication can happen easily if the double strand separates into single strands and the respective nucleotides attach to each strand: "We imagine that prior to duplication the hydrogen bonds are broken and the two chains unwind and separate. Each chain then acts as a template for the formation on to itself of a new companion chain, so that eventually we shall have two pairs of chains, where we only had one before. Moreover the sequence of the pairs of bases will have been duplicated exactly." (WATSON & CRICK 1953, Nature 171, page 966).

The enzyme *primase* synthesizes a short nucleotide sequence called a *primer* to the DNA single strands. The primer is necessary as a starting point for the *DNA polymerase* because these enzymes can indeed elongate a DNA sequence but not start a new one. Individual nucleotides attach to the parental single strands and the respective complementary bases A and T, and C and G bind to each other.

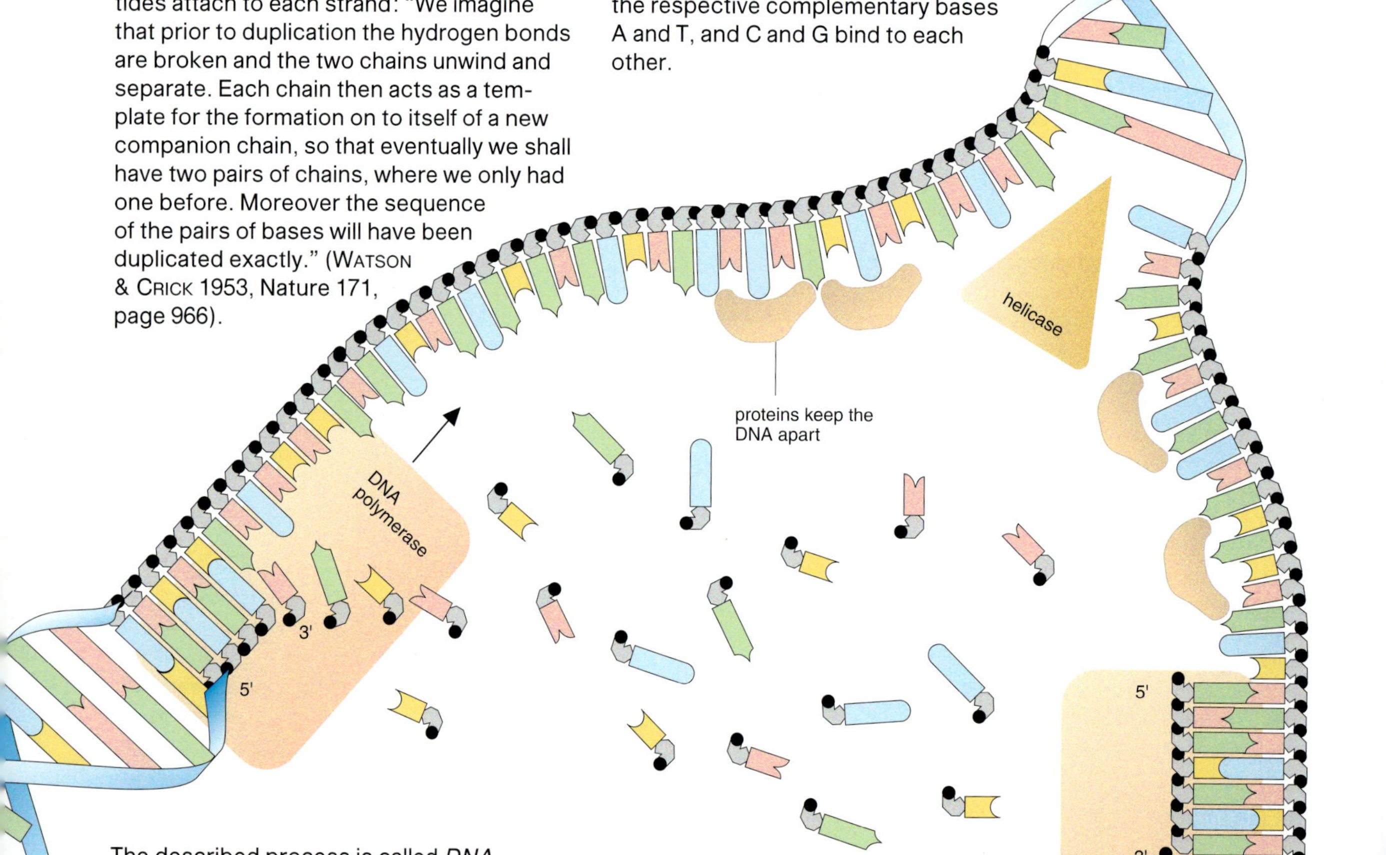

helicase

proteins keep the DNA apart

DNA polymerase

3'

5'

5'

3'

DNA polymerase

The described process is called *DNA replication*. Experiments with DNA from prokaryotes have confirmed the assumption made by WATSON and CRICK and have identified the enzymes that are involved. In the circular chromosome of the bacterium *E. coli*, replication starts at one specific DNA segment (*origin of replication, ori*). Here, a complex consisting of various replication enzymes attaches to the DNA. At this point, the DNA is unwinded and then the hydrogen bonds between the strands are broken by an enzyme called *helicase*. This separates the DNA double strand into single strands. Single-strand-binding proteins attach loosely to the released bases to inhibit the rebinding of the bases. Replication proceeds in two directions from the origin of replication so that two replication forks are formed (see margin p. 13).

Polymerase links the attached nucleotides to a chain. The base sequence of the parental strand is therefore the template that dictates the base sequence of the daughter strand. Hence, the duplicated DNA has the same base sequence as the parental original. Each new double helix consists of one half of the parental *template* strand, the other half being the newly synthesized daughter strand. The replication is thus *semi-conservative.*

On reaching a specific base sequence, the replication enzyme complex detaches from the DNA and replication is hereby terminated. From one DNA ring in *E. coli*, two identical DNA rings have been made.

In contrast to the circular DNA of prokaryotes, the linear DNA of eukaryotes is duplicated in multiple steps and therefore has more than one origin of replication. The replicated segments are linked after their synthesis. Before replication, each *chromosome* consists of one *chromatid* (1 C) and, after replication, it consists of two chromatids (2 C). Both chromatids are attached to each other at the *centromere* and they are not separated until <u>mitosis</u>. Replication takes place during the S-phase of the cell cycle. Prior to that, the necessary enzymes are synthesized in the G1-phase. After S-phase, replication is checked to determine whether it has been successful; accurate replication is essential for the continuing process. Only following a successfully completed replication can the genetic material be transferred in equal parts to the daughter cells during mitosis.

Task

(1) How can the structure of the DNA lead us to the mechanism of replication?

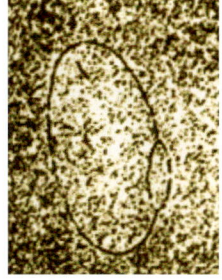

Replication forks in the DNA of *E. coli*

»info box«

From zip to loop

Isolation and analysis of individual replication enzymes have refined the replication model:

The DNA polymerase required for DNA synthesis is a double molecule of multiple parts. It can attach to the two parental single strands like a sliding tab to a zipper. However, it only proceeds in the 3' -> 5' direction and synthesizes the complementary daughter strand from the 5'- to the 3'-end. In the double helix, the single strands are paired with opposite orientation (antiparallel) and only the forward strand template (the 3' -> 5' template strand) can be replicated continuously. Here the *leading strand* is synthesized.

The oppositely orientated template (reverse strand template), however, is replicated in segments, with each segment requiring its own primer. The primer nucleotides are exchanged for DNA nucleotides by the DNA polymerase. The synthesized DNA segments are called after their discoverer *Okazaki fragments*. The fragments are then linked by the enzyme ligase and form the *lagging strand*.

Since the DNA polymerase cannot move at the same time on the forward and on the reverse strand templates in opposite directions, it is assumed that the reverse strand template winds itself around the replication enzyme complex. Within the loop region, the reverse strand DNA is aligned in the same direction as the forward strand template and it can be read by the DNA polymerase piece by piece in the same direction.

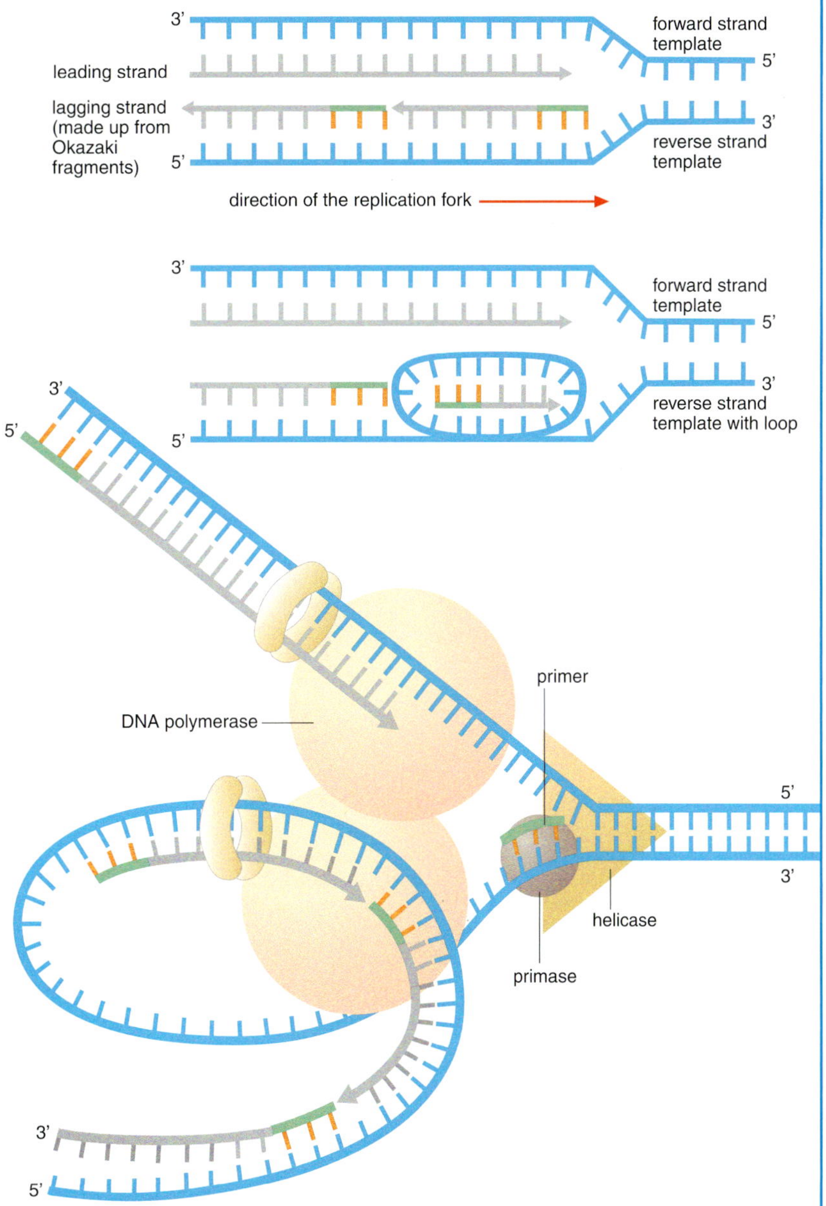

3'

leading strand

lagging strand (made up from Okazaki fragments) 5'

forward strand template 5'

reverse strand template 3'

direction of the replication fork →

3'

forward strand template 5'

reverse strand template with loop 3'

DNA polymerase

primer

5'

3'

helicase

primase

3'

5'

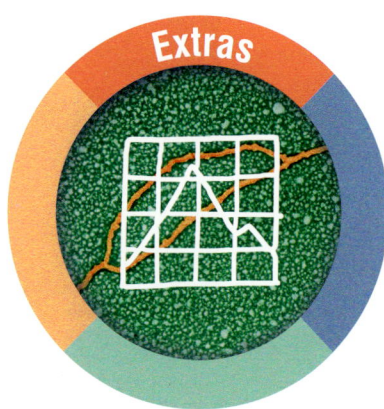

DNA-duplication – how and when?

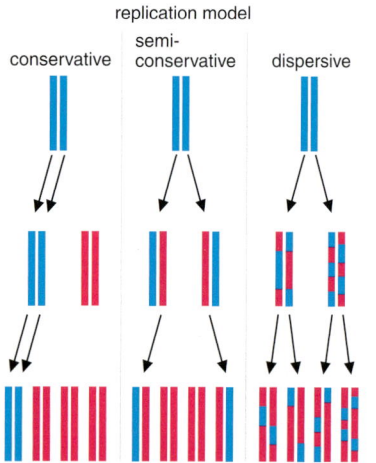

1 Possible replication models

The duplication of the DNA (*replication*) happens before every cell division. In the conservative model, for example, the original could remain unchanged and a copy is made of it in a photocopy-like duplication process. The dispersive replication model (fig.1) is another alternative. On the basis of the DNA structure presented by WATSON and CRICK in 1953, however, *semi-conservative duplication* seems more likely: the DNA double strand is separated into two single strands and, on each single strand, a new complementary strand is synthesized.

The Meselson-Stahl experiment

In 1958, MATTHEW MESELSON and FRANK STAHL used nitrogen isotopes to resolve the DNA replication mechanism. They grew bacteria in culture media that contained the rare heavy ^{15}N (not radio-active) instead of the naturally occurring light ^{14}N nitrogen isotope. Bacteria cannot distinguish between the two isotopes and incorporate the tagged heavy nitrogen into their cellular components and also into their DNA. DNA containing e.g. ^{15}N instead of ^{14}N within the organic bases has a greater density ("heavy DNA").

MESELSON and STAHL grew an *E. coli* culture in the presence of ^{15}N. They then changed the culture medium and only ^{14}N was made available. The newly synthesized DNA therefore only contained ^{14}N. The bacteria were given enough time to divide once, a process that takes about 20 min in *E. coli*. A few cells were removed from the culture in order to extract and purify their DNA. After a second division cycle, the bacteria were again removed from the culture and their DNA was also isolated.

In order to distinguish between "light" and "heavy" DNA, the researchers invented a method called *caesium chloride equilibrium centrifugation*. The isolation method is based on the theory that a particle floats in a solution when its density equals the density of the surround-

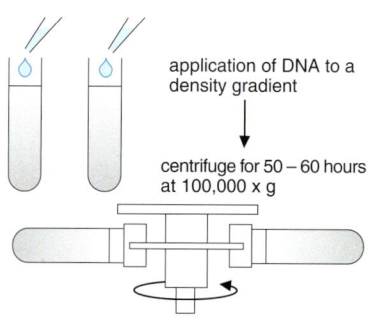

application of DNA to a density gradient

centrifuge for 50 – 60 hours at 100,000 x g

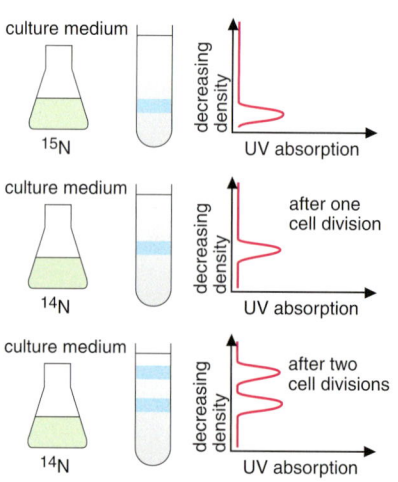

2 Results obtained by MESELSON and STAHL

ing fluid. In order to create a concentration gradient and thereby a density gradient, a caesium chloride solution is exposed to an extremely strong artificial gravity field in an ultracentrifuge. In the bottom part of the test tube, a higher caesium chloride concentration forms with a greater density than that in the top part of the test tube (see fig. 2). If a DNA sample is applied to this gradient and centrifuged again, the DNA migrates within the density gradient until its density equals the density of the caesium chloride solution (see fig. 2). Since DNA is not visible to the naked eye but absorbs UV light, its position can be identified by detecting the UV light absorption.

Tasks

① DNA is replicated by a semi-conservative mechanism. Explain using figure 2.
② What density distribution would you expect for conservative or dispersive duplication?

Autoradiography

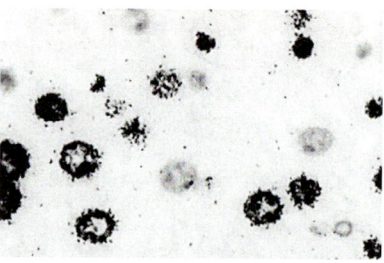

The number of cells in mitosis is detectable by light microscopy. To determine the number of cells in S-phase, a special marker is needed. The cells are exposed to radioactively marked DNA nucleosides, most commonly [^3H]-thymidine. Only DNA replicating cells incorporate this molecule into their DNA. Autoradiography is used to detect the cells thus marked: the cells are washed to remove free [^3H]-thymidine, fixed and then covered with a photographic film. Radioactive decay causes a blackening of the film.

Task

③ The figure shows an autoradiograph of liver cells. How high is the percentage of replicating cells?

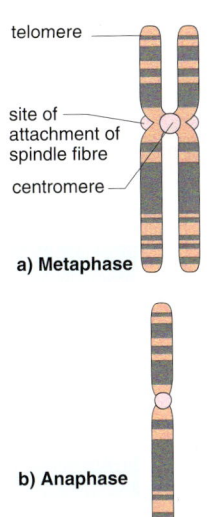

telomere

site of attachment of spindle fibre

centromere

a) Metaphase

b) Anaphase

c) G$_0$- and G$_1$-phase

d) G$_2$-phase

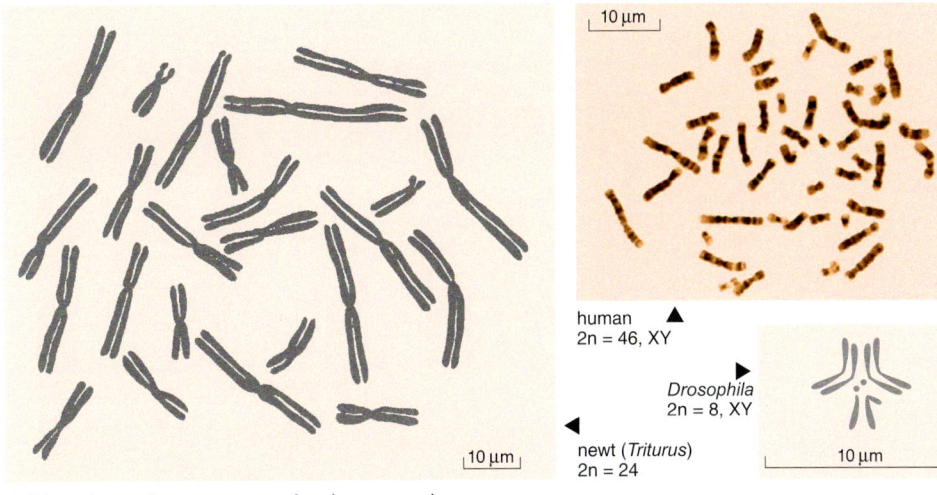

human
2n = 46, XY

Drosophila
2n = 8, XY

newt (*Triturus*)
2n = 24

10 µm

10 µm

10 µm

1 Metaphase chromosomes of various organisms

Karyograms show chromosomes in metaphase

As early as the middle of the 19th century, thread-like structures were detected in the nuclei of dividing cells. The German Professor of Anatomy WILHELM WALDEYER called them *chromosomes* (Greek: *chroma* = colour, *soma* = body). Chromosomes can be stained with special dyes, then they often express a characteristic type of banding.

With a light microscope, chromosomes can best be identified during the metaphase of mitosis. At this stage, the cell has passed through the S-phase and therefore all *metaphase chromosomes* contain two genetically identical chromatids (2C) that are still attached at the *centromere*. This results in an X- or V-shape. The centromere is the binding site of proteins to which spindle fibres attach. The ends of the chromosomes show specific structures called telomeres, which for example inhibit adhesion between chromosomes.

Metaphase chromosomes (fig. 1) can be arranged in a *karyogram* according to their size, shape and banding pattern. In order to do this, cell division is stopped in metaphase by a toxin called *colchicine*, which affects the spindle apparatus. The chromosomes are stained, micrographs (pictures) are taken via a light microscope and then the chromosomes are sorted. Karyograms show that chromosomes in all somatic cells have an identically shaped partner. The chromosomes of one pair resemble each other in shape but genetically they are often not identical: In somatic cells, two sets of homologous chromosomes are present with each set containing n chromosomes. Such cells are named *diploid* (Greek: *diplos* = double). The number and shape of the chromosomes in diploid cells (2n) are species-specific. However, there may be differences between male and female cells of the same species; these differences are usually restricted to one chromosome pair. Such chromosomes are called *gonosomes* (*sex chromosomes*); the remaining chromosomes are called *autosomes*.

Humans possess 22 pairs of autosomes and the gonosomes X and Y. A karyogram can also be briefly represented by its karyotype. In a *karyotype*, first the number of all chromosomes is stated and then the types of gonosomes. The female human karyotype is thus 46, XX; the male human karyotype is 46, XY.

During the interphase of the cell cycle, the chromosomes are so loosely packed that we are only able to see them as chromatin under the light microscope. However, the nucleus still contains two homologous sets (2n) of *interphase chromosomes* in the form of loosely coiled threads of DNA. Prior to replication in the S-phase, each chromosome consists of one chromatid (1C) and, after S-phase, of two chromatids (2C).

The term "chromosome" is nowadays no longer restricted to eukaryotes but is also used for the circular DNA unit of prokaryotes.

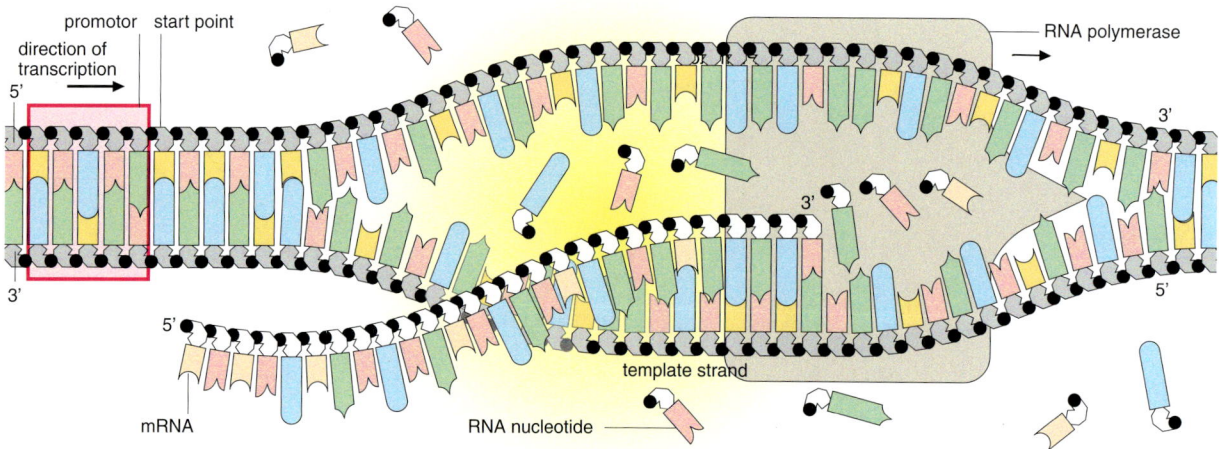

1 Transcription: mRNA is synthesized by a DNA-dependent RNA polymerase

Gene expression: from information to product

DNA duplication and mitosis make sure that all cells of an organism contain the same genetic information. Despite this, mammals for example consist of more than a hundred different cell types. Since proteins are essential for cellular structure and function, the different cell types also vary in their protein content. For example, muscle cells have a high concentration of motor proteins such as actin and myosin. By electron microscopy the highly organized structure causing the cells to contract is visible. Antibody-producing immune cells, however, do not have this structure. Even during the life of a cell, its protein pattern changes depending on whether the cell is preparing for division or is not able to divide.

Gene expression is the process in which a cell gains access to the information held by its genes. It can be subdivided in two steps:

1. *Transcription*: an RNA copy is made from the DNA template.
2. *Translation*: a polypeptide is synthesized by ribosomes according to the RNA template.

Transcription: genetic information on the move

The steps leading from a gene to a protein comprise, as a first step, the rewriting of a gene from DNA to RNA (ribonucleic acid) in a process called *transcription* (Latin: *transcribere* = to rewrite). Genetic information thus becomes moveable and can leave

the nucleus of eukaryotic cells; it can be transported from the DNA to the ribosomes, which are the organelles of protein biosynthesis. There are several different types of RNA. The RNA that transfers the genetic message is called messenger RNA (mRNA). RNA differs from DNA on some points:

— It is made up of only one polynucleotide chain, whereas DNA consists of two polynucleotide strands.
— It is much shorter because it only represents a small segment of the genomic information.

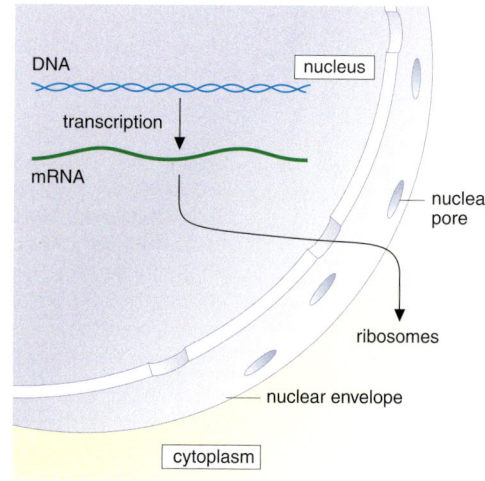

2 The genetic information leaves the nucleus

DNA nucleotide with thymine

RNA nucleotide with uracil

Gene
A DNA segment that codes for an RNA (see page 26)

Transcription
Production of a complementary RNA molecule from a DNA template strand in the G-phase of the cell cycle

RNA
Ribonucleic acid

- It contains the pentose sugar ribose instead of deoxyribose (therefore the abbreviation RNA).
- Instead of the base thymine, the similarly built base uracil, which also pairs with adenine, is present in RNA.

Transcription resembles DNA replication to a certain extent. The DNA unwinds and is separated into two single strands. Complementary nucleotides attach to the template strand according to the base-pairing rule.

They are then linked by the enzyme RNA polymerase to form an RNA single strand called mRNA. The RNA polymerase, however, requires information about which DNA single strand should be transcribed and in which direction transcription should proceed. For this purpose, special DNA regulatory sequences called _promoters_ are located in front of the affected DNA segment. They are the starting point and, at the same time, also indicate the direction of transcription. Since RNA polymerases can also only synthesize RNA in 5' to 3' direction, it is thereby predetermined which DNA single strand will function as the _template strand_: the forward strand template running from 3' to 5'.

Transcription terminates when the RNA polymerase reaches a specific DNA base sequence on the DNA template strand called the _terminator_. The enzyme then releases the template strand and the mRNA.

Task

① Make a table and compare DNA replication and transcription regarding function, time point in the cell cycle, respective templates, involved enzymes and used nucleotides.

»info box«

Giant chromosomes: Transcription made visible

In 1881, "nuclear threads" in the salivary gland cells of chironomids caught E.-G. BALBIANI's eye. Decades later, these threads were identified as _giant chromosomes_. Giant chromosomes form in nuclei that perform replication but no subsequent mitosis. The DNA is duplicated but not distributed among daughter cells. Each further replication doubles the number of chromatids. After r replications, every chromosome is made up of 2^r identical chromatids. Often, not only the identical chromatids but also the homologous chromosomes are paired. This is the reason that we can only count n giant chromosomes by light microscopy. Giant chromosomes have a typical banding pattern because the DNA segments stain differently and the individual chromatids lie exactly next to each other. Additional to the typical banding, we can see parts that are "puffed up", the so-called _chromosome puffs_.

On these less condensed DNA segments, transcription takes place in many chromatids at the same time and is thus highly effective. This has been established by radioactively marked RNA nucleotides accumulating at the puffs. Since different proteins are differentially required in the various tissues and development stages of a chironomid larva, different genes are active and therefore variable patterns of puff activity are expressed. Giant chromosomes must be interphase chromosomes, as transcription and replication take place there.

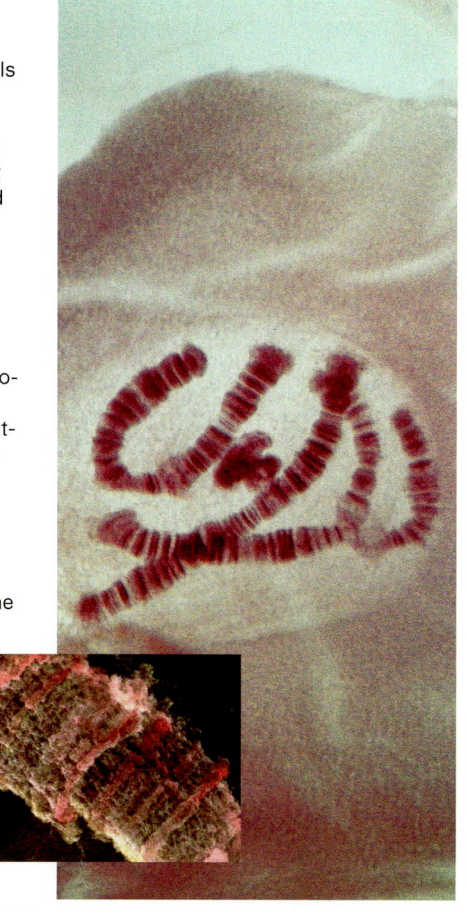

The genetic code

The structure and function of proteins are determined by the order of their amino acids: their *amino acid sequence*. Theoretically, an almost unlimited number of amino acids sequences can be made based on 20 different amino acids. The base sequence of the mRNA determines in which order the amino acids have to be linked to form long chains making a functional protein.

A code can be used to translate signals of one system to signals of another system, e.g. from the Morse alphabet to common letters. In the same way, the genetic code provides the signals for the translation of an mRNA base sequence to the amino acid sequence of a protein. The nucleotide sequence of the mRNA codes for the amino acid sequence.

Since there are only 4 different bases but 20 protein-forming amino acids, a group of multiple nucleotides must form a code word (*codon*) coding for one amino acid. If one base stands for one amino acid, four bases can represent only 4 different amino acids. Units of two bases can represent only 4^2 = 16 amino acids but 3 bases in a unit can code for 4^3 = 64 different amino acids.

A group of three, a *base triplet*, is therefore the theoretical minimum for a codon. However, then there are more base triplets than coded amino acids. Some amino acids are encoded by several base triplets. The genetic code is thus *redundant*.

The length and meaning of the individual mRNA codons has been determined experimentally by using artificial mRNA molecules with a known base sequence and by analysing the subsequently synthesized amino acid chains. The genetic code is indeed a triplet code. It is redundant because most amino acids are encoded by several different codons that mostly differ only in their third base. The rare amino acids methionine and tryptophan are each represented only by one codon. Other codons have special meanings similar to our punctuation marks: the *start codon* AUG represents the DNA starting point of transcription and its respective amino acid (an altered methionine) is later removed from the protein. UAG, UAA and UGA do not represent any amino acid but mark the termination of the translation process. They are therefore called *stop codons* (or termination codons). In almost all organisms, the codons have the same meanings and therefore a given mRNA is translated into the same amino acid sequence. Thus, the genetic code is universal.

Colinearity
The sequence of nucleotides in the mRNA corresponds to the sequence of amino acids in the coded protein

The **codon wheel**
shows which mRNA codon is translated to which amino acid. The first nucleotide of a codon (5'-end) is located inside — the codons are read from inside out. GCA for example stands for the amino acid alanine

Tasks

1. Look for the start codon and translate the following mRNA sequence: 5'UUAGAUGAGCGACGAACCC-CUAAAAUUUACCUAGUAGUAGC-CAU3'

2. Into which amino acid sequence is the following segment of the DNA template strand translated? 3'CTGGCTACTGACCCGCTTCTTCTATC5'

3. When we remove the first letter of the example sentence "thebigfoxbitthedog" and keep to the "triplet code", the meaning of the sentence is lost. What happens if we remove the first base of the DNA sequence described above?

1 Codon wheel

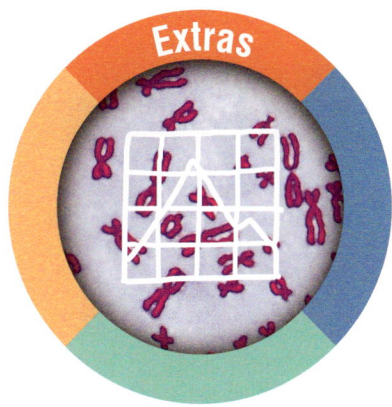

The discovery of the genetic code

For a long time, it was unclear which base triplet coded for which amino acid. In 1964, the researchers MARSHALL NIRENBERG and PHILIP LEDERER described an experimental approach that made it possible to resolve the genetic code completely during the subsequent years.

Testing triplet binding

NIRENBERG and LEDERER synthesized short mRNA molecules of known base sequence. They were mixed with ribosomes isolated from bacteria. The experiments showed at first that the mRNA molecules bound to the ribosomes. In a second step, they added to a test tube all components necessary for protein biosynthesis. This cell-free system mainly contained:
— purified ribosomes
— a mixture of all 20 amino acids present in organisms (alternating from experiment to experiment which amino acid was radioactively marked)
— an mRNA with three nucleotides of known base sequence

After the mentioned components had been mixed and could react with each other, the mixture was applied to a filter. The size of the filter pores was chosen to retain particles of the size of ribosomes and to allow smaller particles to pass through. The radioactive signal was then studied by testing its presence on the filter and in the filtrate.

Four key experiments

The two researchers added mRNA with the base sequence UUU to their cell-free system. If they marked the amino acid serine radioactively, the radioactive signal was found in the filtrate. If phenylalanine instead of serine was marked radioactively, the radioactive signal was found on the filter. In a second experimental set, they used mRNA with the base sequence UCU. Radioactively marked serine was found on the filter and radioactively marked phenylalanine was found in the filtrate.

Tasks

① Make an outline of the experimental approach employed by NIRENBERG and LEDERER to decode the genetic code.
② What information could be drawn from the experimental results concerning the triplets UUU and UCU?
③ All mRNA molecules of the original cells were removed from the ribosome samples prior to the triplet binding tests. Explain why.

Further experiments

When longer mRNA molecules of known base sequence are added to the cell-free system, polypeptides based on the information in the artificial mRNA are synthesized. The amino acid sequence of the isolated polypeptides can then be determined.

With this method, M. NIRENBERG and H.G. KHORANA were able to find more triplets. The table shows some of the results of these experiments. The synthesized mRNA molecules are shown left, with the forming peptides being given right. The RNA is abbreviated: since they are repetitive polynucleotides, it is sufficient to indicate the repetitive segments. Therefore, poly-U means an mRNA molecule made only from uracil nucleotides (-UUUUUUUUU-), poly-A is an mRNA made from only adenine nucleotides (-AAAAAAAAAA-) and, in poly-AC, adenine and cytosine nucleotides alternate regularly (-ACACACACACACA-CAC-).

Tasks

④ When using poly-U, poly-A, poly-C or poly-G respectively, the polypeptides shown below will be obtained, containing only one type of amino acid. This shows the meaning of 4 of the triplets. Which are they?
⑤ When using mRNA with alternating nucleotides, peptides are obtained in which two amino acids alternate. Explain this. Is the meaning of the triplets clearly solved by these experiments?
⑥ When we use other regular polynucleotides with longer subunits, we obtain a mixture of different peptides (see below). Why are various different peptides formed? Use the information from task 4) and 5) and explain the meaning of further nucleotides.
⑦ Even mRNA molecules with 4 regularly alternating nucleotides were constructed. When we look at the product (see table), which triplets can we explain? Why does the primary structure show repetition after 4 amino acids?

RNA	The forming peptides
Poly-U	Phe - Phe - Phe - Phe - Phe - ...
Poly-A	Lys - Lys - Lys - Lys - Lys - Lys - ...
Poly-C	Pro - Pro - Pro - Pro - Pro - Pro - ...
Poly-G	Gly - Gly - Gly - Gly - Gly - Gly - ...
Poly-AC	Thr - His - Thr - His - Thr - His - ... or His - Thr - His - Thr - His- Thr - ...
Poly-AAC	Asn - Asn - Asn - Asn - Asn ... or Thr - Thr - Thr - Thr - Thr - Thr ... or Gln - Gln - Gln - Gln - Gln - Gln - ...
Poly-ACC	Thr - Thr - Thr - Thr - Thr - ... or Pro - Pro - Pro - Pro - Pro - ... or His - His - His - His - ...
Poly-ACCC	Thr - His - Pro - Pro - Thr - His - Pro - Pro - ...

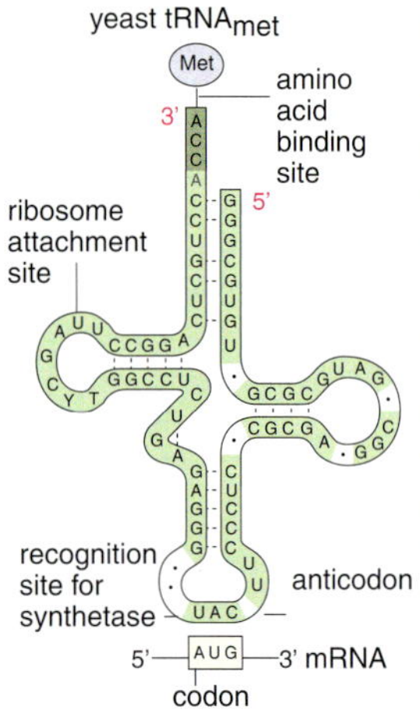

tRNA – mediator between mRNA and peptides

yeast tRNA_met

ribosome attachment site

recognition site for synthetase

anticodon

5' AUG 3' mRNA

codon

amino acid binding site

The genetic code indicates which base triplet is translated into which amino acid in the respective polypeptide. It represents the dictionary for translation. A special nucleic acid called *transfer RNA* (tRNA; Latin: *transferre*) works as a mediator.

It is the link between the base and amino acid sequence. The tRNA consists of ribonucleotides and is cloverleaf-shaped. The tRNA molecule has two exposed binding sites:

— The so called *anticodon*, which binds complementarily to specific mRNA triplets according to the base-pairing rule.
— The amino-acid-binding site to which one specific amino acid attaches. This amino acid is the one that is encoded by the respective mRNA triplet according to the genetic code.

Highly specific enzymes, the so-called *aminoacyl-tRNA synthetases*, recognize the anticodon of a tRNA and load it with the respective amino acid. One can also say that the genetic code depends on the ability of the aminoacyl-tRNA synthetases to assign specific RNA triplets to defined amino acids. This is the crucial step during the realisation of the genetic code. In all cells, 20 different synthetases are present, each being responsible for one amino acid only. About 50 different types of tRNA are known, more than enough to bind one of the 20 amino acids specifically. Therefore, more than one tRNA molecule can exist per amino acid. Most amino acids are encoded by numerous triplets since the genetic code is redundant. Transfer RNA molecules are also encoded by DNA segments.

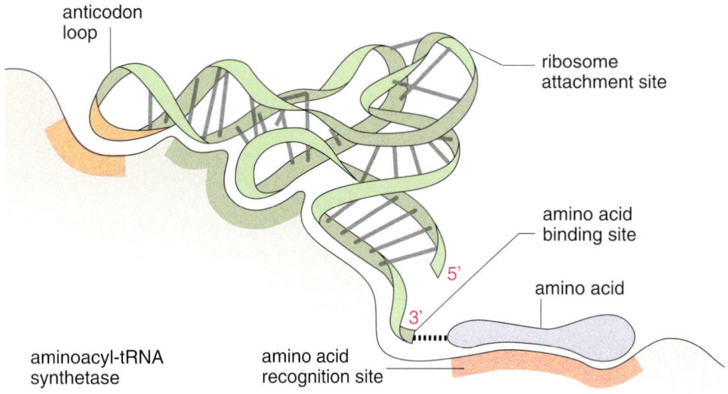

anticodon loop

ribosome attachment site

amino acid binding site

amino acid

aminoacyl-tRNA synthetase

amino acid recognition site

1 Loading of a tRNA with an amino acid

»info box«

Wobble hypothesis

We might expect that each cell contains 61 different tRNA molecules, one for every mRNA codon coding for an amino acid. However, we know that one cell has no more than 31 different tRNA molecules. F. CRICK postulated the *wobble hypothesis*.

Based on the spatial structure of tRNA, unusual base pairing is possible within the short complementary pairing part of the anticodon and codon. A single anticodon can therefore pair with more than one codon type. Codons read by the same type of tRNA form a codon family. The rules of the genetic code remain uncorrupted — the code does not wobble.

The unusual base inosine (I) can pair with the bases C, A and U. AUC, AUU, AUA code for the amino acid isoleucine.

Codon preference

Surprisingly, codons with the same meaning are not always used with the same frequency. In systematically closely related groups of organisms, often the same codons are preferred and others are avoided. To some extent, different groups of organisms use diverse genetic dialects.

The four triplets CCC, CCG, CCA and CCU code for the amino acid proline. In the humane genome, the triplet CCC codes for 33 % of all proline molecules. It is therefore the preferred triplet. In the rat, the preferred triplet is CCC (32 %), in *Escherichia coli* it is CCG (51 %), in the yeast *Saccharomyces cerevisiae* CCA (41 %) and in *Saccharomyces frugiperda* the triplets CCA and CCT are used with the same frequency of 29 %.

Task

① Codon preference is mirrored in the concentration of tRNA molecules.
What problem might occur if a gene from, for example, *Escherichia coli* is transferred to *Saccharomyces frugiperda* in order to produce the particular protein?

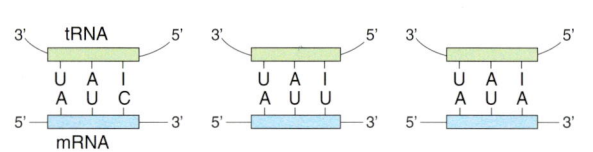

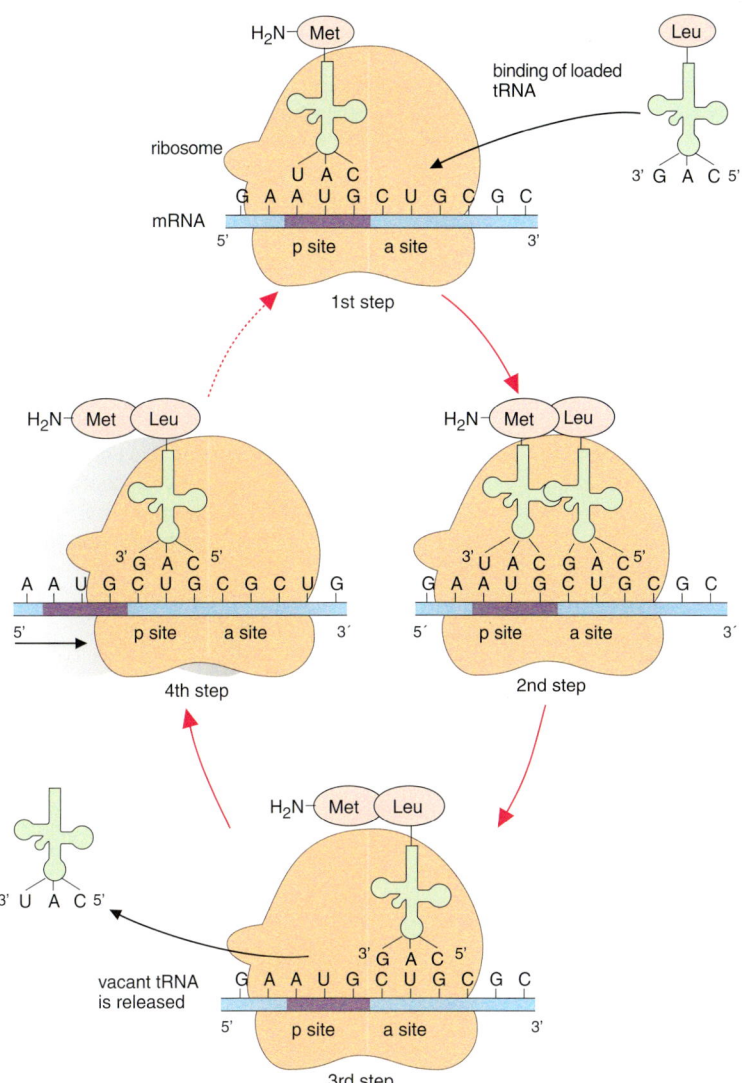

1 Translation process: elongation of the amino acid chain

base pairing, methionine-tRNA attaches with its anticodon (UAC) to the start codon (AUG). Additionally, the tRNA molecule binds to the P site of the ribosome. Another loaded tRNA with the respective anticodon for the next codon on the mRNA binds to the A site (1st step, fig. 1). If the codon and anticodon are complementary, the respective tRNA is strongly bound. Non-complementary molecules detach. If the bases are complementary, the amino acids of both tRNA molecules are chemically linked using energy (2nd step). The ribosome is responsible for this enzymatic reaction.

In order to elongate the amino acid chain, the ribosome is moved in the direction of translation (5' to 3') by one base triplet on the mRNA. The tRNA from the P site moves out of the ribosome and detaches (3rd step). The remaining tRNA moves from the A site to the P site so that the A site is vacant again (4th step). The A site is now located adjacent to the next codon of the mRNA and can suitably bind a tRNA with the complementary anticodon. The respective amino acid is linked with the dipeptide forming a tripeptide. The peptide is in this way continuously elongated by one amino acid.

The elongation process is repeated until the ribosome reaches a *stop codon* on the mRNA. According to the rules of the genetic code, there is no corresponding tRNA to a stop codon. Translation termination begins: the complex of the ribosome and RNA separates and the synthesized amino acid chain is released.

Translation: a protein develops

Translation is the second step after transcription during gene expression. In this step, the *base sequence* of the mRNA is translated into the amino acid sequence of the protein. Translation occurs on the *ribosomes*. Ribosomes are the organelles of protein biosynthesis and consist of protein and the nucleic acid called rRNA. A functional ribosome consists of two subunits of different sizes.

Translation is organized into four steps. To begin with, the two ribosomal subunits assemble at the *start codon* of the mRNA. According to the rule of complementary

Formation of peptide bond

Tasks

① RNA and proteins belong to chemically different classes of substances. However, they exhibit structural and functional similarities. Explain why.

② The codon AUG has two meanings depending on whether it is located at the beginning or in the middle of a mRNA. Explain why.

③ Find the complementary mRNA to the DNA template strand below. Then find the complementary anticodons and the respective amino acids:
3'CTGGCTTGAACCCGCTTC5'

Protein biosynthesis in prokaryotes

Prokaryotes
Organisms such as bacteria containing neither nucleus nor membrane-enclosed organelles

Eukaryotes
Organisms such as plants and animals containing a nucleus and membrane-enclosed organelles

The basics of transcription and translation were first examined closely in *prokaryotes*. The *promoter* (the recognition and binding site of RNA polymerase) was shown to dictate the reading direction of the polymerase. On the 3' to 5' DNA single strand, complementary RNA nucleotides are attached with a velocity of about 2,500 nucleotides per minute. This DNA strand represents the template strand and dictates the "meaning", i.e. the base sequence, of the mRNA. The base sequence of the forming mRNA can easily be predicted from the opposite DNA strand just by replacing T with U. The mRNA in prokaryotes is readily accessible during its synthesis. Ribosomes can attach to the free 5'-end, which forms first, even while transcription of the following segments is still continuing. If the first ribosome moves forward on the mRNA, the next ribosome can attach. In this way, the mRNA is read simultaneously by many ribosomes. This complex is called a *polysome*.

In a bacterial cell, newly synthesized protein molecules can be detected only half a minute after the start of the transcription. However, the used mRNA is also quickly degraded after a few minutes. This short life span is biologically significant for bacteria because of their short generation interval of less than 30 minutes under optimal conditions.

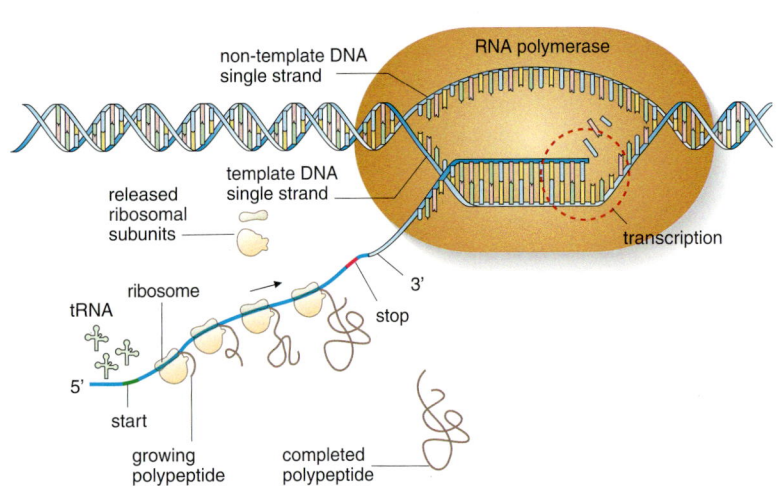

1 Summary of protein biosynthesis in prokaryotes

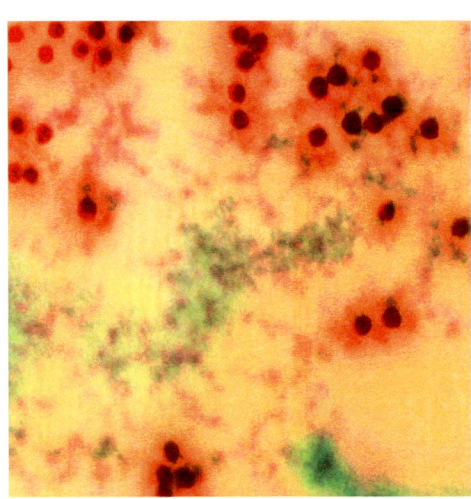

2 Polysomes

»info box«

DNA in chloroplasts and mitochondria

Chloroplasts and mitochondria are not produced by cells but they reproduce by division. Both cell organelles contain their own DNA, which is different from the DNA of the nucleus, e.g. it has no introns (see page 23). The mitochondrial DNA is haploid (no recombination) and relatively small. The DNA is circular, as in prokaryotes, and is not associated with histones. Mitochondrial genes are transcribed inside the mitochondrion and are translated on its own ribosomes, which resemble bacterial ribosomes. Mitochondrial DNA codes for some enzymes of energy metabolism as well as for tRNA and rRNA molecules.

The genome of chloroplasts is also circular and contains up to 120 genes. The ribosomes of chloroplasts also resemble bacterial ribosomes. Based on these similarities, chloroplasts and mitochondria in eukaryotes are assumed to have evolved from absorbed prokaryotes (*endosymbiotic hypothesis*).

Many different mRNA molecules can function in a short period of time and the synthesized proteins can immediately control various metabolic and developmental processes. Additionally, a fast response to changing environmental conditions is possible because the required proteins can be produced quickly and their synthesis can be stopped as soon as the present demand is satisfied. Nucleotides of the degraded mRNA will be reused.

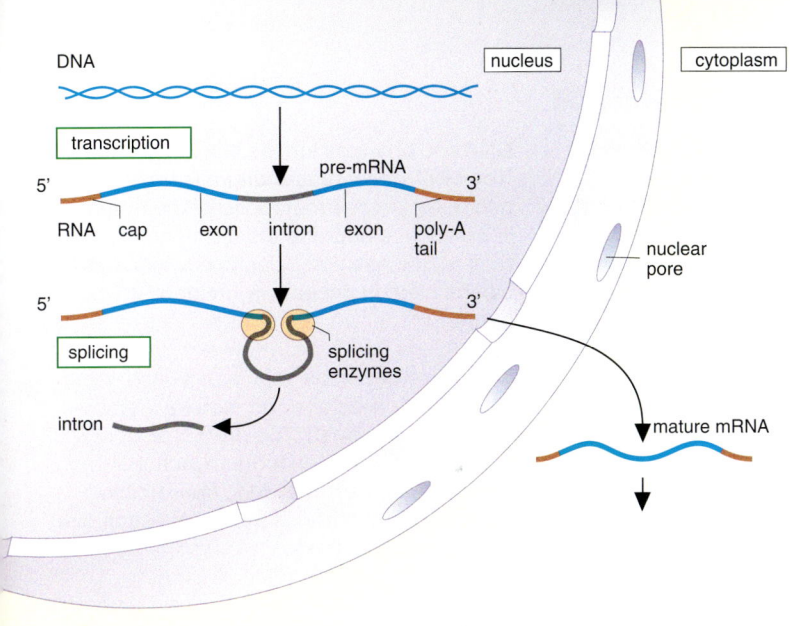

DNA | nucleus | cytoplasm

transcription

pre-mRNA

5' — 3'

RNA | cap | exon | intron | exon | poly-A tail

nuclear pore

5' — 3'

splicing — splicing enzymes

intron

mature mRNA

1 Splicing

any information; these are called *introns* and can be between 50 and 30,000 nucleotides long. Cutting enzymes place the introns at specific sites in a loop and then cut them out. The exons are linked to form the actual mRNA. The entire process is called *splicing*. Prokaryotes have neither introns nor splicing processes.

Eukaryotic mRNA molecules have a considerably longer life span than prokaryotic ones. After transcription, a protective group composed of 150-200 adenine nucleotides *(poly-A tail)* is attached to the 3'-end of eukaryotic mRNA. This slows down enzymatic degradation considerably. A "cap" of methylated guanine is attached to the 5'-end. This is important for the contact with the smaller ribosomal subunit during the start of the translation. The mature *mRNA* is transported from the nucleus through nuclear pores to the *rough endoplasmic reticulum (rER)*. Translation takes place on the ribosomes of the rER.

Many proteins are still modified after their synthesis at the ribosomes. Pre-mature insulin, for example, contains two amino acid segments that are not present in the end-product insulin found in body fluids. The *pre-sequence* makes the passage into the endoplasmic reticulum (ER) possible. Here, it is removed. The *pro-sequence* is also removed inside the ER. Only this end-product can bind to the insulin receptor. Digestive enzymes of lysosomes are marked by a mannose phosphate residue and, because of this marker, they can be packed into lysosomes. If this sugar residue is missing because of genetic defects, the enzymes pass into the body fluids.

Protein biosynthesis in eukaryotes

Protein biosynthesis in eukaryotes basically proceeds along the same steps as in prokaryotes. However, one striking finding is that the RNA initially produced by transcription is significantly longer than is required by the coded protein. It is a "precursor" (pre-mRNA) and has to be converted to a functional mRNA before translation. Only a few segments of the RNA (*exons*) contain information about the protein. In between, there are segments that (seemingly) do not contain

Tasks

① Human insulin can be produced in "reprogrammed" bacteria. What problems might be expected because of the different genetics in pro- and eukaryotes?

② Inhibitors block specific steps. In order to examine protein biosynthesis the use of such inhibitors has been extremely important. Explain which step is blocked by the inhibitors described in the table. Which inhibitor could theoretically be used as an antibiotic in medicine?

Inhibitor	Effect on protein biosynthesis
Actinomycin D	binds only to prokaryotic DNA
Rifampicin	binds to the RNA polymerase of prokaryotes
Amanitin	binds to the RNA polymerase of eukaryotes
Streptomycin	binds to the small ribosomal subunit of prokaryotes
Tetracycline	inhibits the binding of tRNA to bacterial ribosomes
Puromycin	binds strongly to the ribosome instead of a loaded tRNA
Chloramphenicol, Erythromycin	binds to bacterial ribosomes
Kirromycin	prevents movement of the tRNA from the A to the P site in prokaryotes
Cycloheximid	prevents the linkage of amino acids on eukaryotic ribosomes

The regulation of gene expression

lactose

tryptophan

Lactose and tryptophan

Many proteins are constantly needed in the cells, e.g. enzymes of energy metabolism or structure-forming proteins of the cytoskeleton. Other proteins are only necessary under special conditions, e.g. in bacteria, if lactose is offered as a nutritional sugar instead of glucose. Genes being constantly transcribed are called _constitutive genes_. Genes that are rapidly turned on or off when required are called _facultative genes_.

The operon model

FRANÇOIS JACOB and JACQUES MONOD examined, in bacteria, the mechanism that is responsible for turning genes on or off. They proposed the operon model for the regulation of gene activity and received, together with ANDRE LWOFF, the Nobel Prize in 1965. Their proposals for the regulation of gene activity at the level of transcription were later confirmed by methods of molecular biology. The model is also valid for eukaryotic genes.

Substrate induction

A repressor protein is constantly produced by a regulatory gene. When no lactose (milk sugar) is present, the repressor binds to a specific DNA segment called an _operator_ and thus blocks the transcription of the genes that are needed for lactose metabolism (fig. 1, top). If a lactose molecule enters the cell, it acts as an _effector_ and binds to the repressor (fig. 1, bottom). The repressor changes its structure, detaches from the

DNA and changes into its inactivated state. Now RNA polymerase can transcribe the genes for lactose metabolism. The translated enzymes degrade the disaccharide lactose to glucose and galactose, which then can be used in the energy metabolism of the bacterium.

The operator can be seen as a switch deciding whether the genes for lactose metabolism are transcribed. The promoter, operator and controlled genes form a functional unit called the _operon_. In the lac operon, the substrate (lactose) has the effect of inducing the synthesis pathway.

End product repression

In end product repression, the high concentration of the end-product causes the synthesis pathway to be turned off (_negative feedback_). The synthesis of the amino acid tryptophan is such a process with negative feedback: a regulatory gene produces a repressor that is inactive — it does not bind to the operator in the DNA.

The genes coding for the enzymes of tryptophan metabolism are transcribed to mRNA without obstruction and are then translated (fig. 25.1, top). The tryptophan concentration rises inside the cell and the chance that a tryptophan molecule can come into contact with the repressor protein and bind to it steadily increases (fig. 25.1, bottom). The repressor then changes its spatial structure, becomes active and attaches to the operator. The tryptophan operon is blocked and the transcription of the tryptophan gene is stopped.

Tryptophan synthesis is repressed. A decreasing tryptophan concentration causes the detachment of the tryptophan molecule from the repressor protein. The inhibition is reversible. This is biologically reasonable because bacteria can thus react flexibly and adapt to changing environmental conditions.

Transcription activators

Transcription activators are proteins that specifically increase the transcription of a gene. The heavy metal copper is a trace element that is needed in low concentrations in all cells. Too a high concentration, however, can severely damage the cell and

1 Substrate induction: the lac operon of E. coli

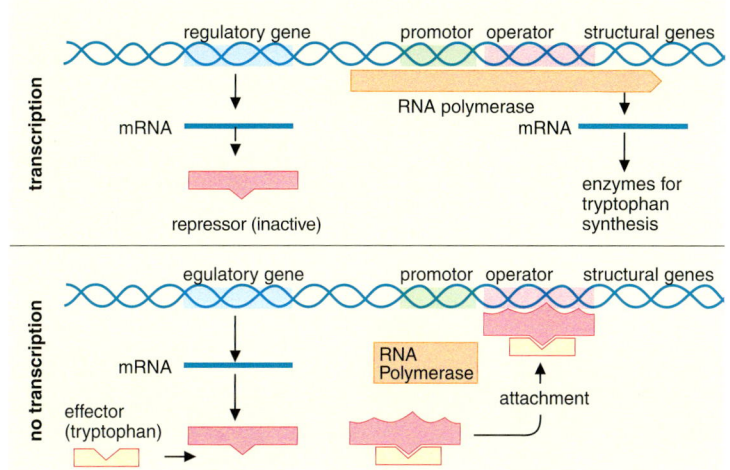

1 End product repression: the tryptophan operon of *E. coli*

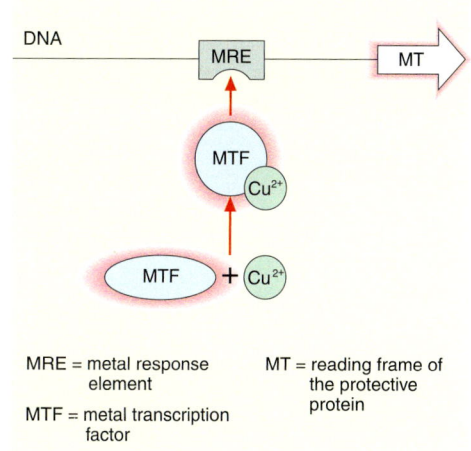

MRE = metal response element

MTF = metal transcription factor

MT = reading frame of the protective protein

2 Gene induction in yeast cells

even cause cell death. Yeast cells react to high copper concentrations by synthesizing protective proteins that bind to copper and that dispose the ions. If copper ions enter a yeast cell, they bind to a transcription factor (MTF) and activate it. This complex binds to the regulatory sequence (MRE) of the gene (MT) coding for the protective molecule. Polymerase recognizes this complex and transcribes the gene.

»info box«

Hormone induction

The largest part of a seed is the endosperm. It consists of the reserve substance starch and dead cells. It is surrounded by the *aleurone layer* consisting of living cells. Starch is used as a nutrient for the embryo during germination. If the seed is placed in water, it swells and the amylase gene in the aleurone cells is activated. After some time, the enzyme amylase is released into the endosperm. This enzyme degrades starch to sugar, which is transported to the embryo. The expression of the amylase gene is caused by the hormone *gibberellin*, which is released from the embryo.

An experiment on gene regulation in wheat seeds

Soluble starch (100 mg) and agar (1 g) are boiled in water (100 ml). The liquid agar is poured into a petri dish. Wheat seeds are left in water for three hours. The swollen seeds are then cut crosswise through the middle and, with the cut surface down, both halves are placed onto the starch agar. After 24 hours, the agar is coated with diluted iodine-potassium solution to detect starch.

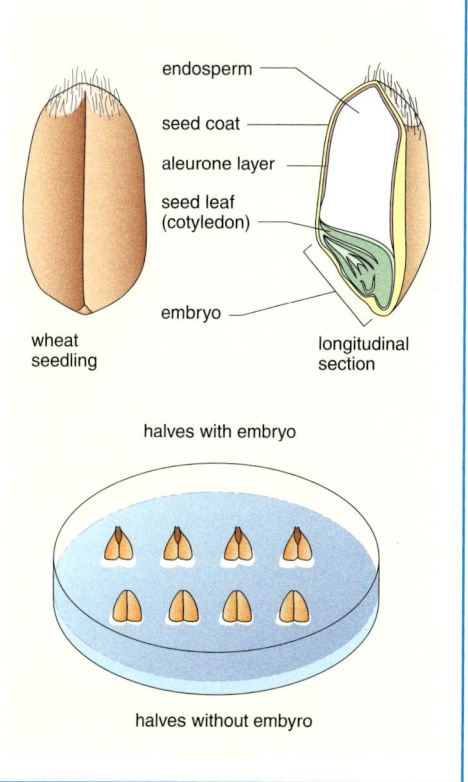

Tasks

① How do you explain the decolouration of the agar?
② Why have only 50% of the halves formed decolouration zones?
③ What result would you expect from wheat seeds from which the embryos have been removed even before the experiment started?
④ The hormone gibberellin can be synthesized artificially. What result would you expect if gibberellin is added to the starch agar?

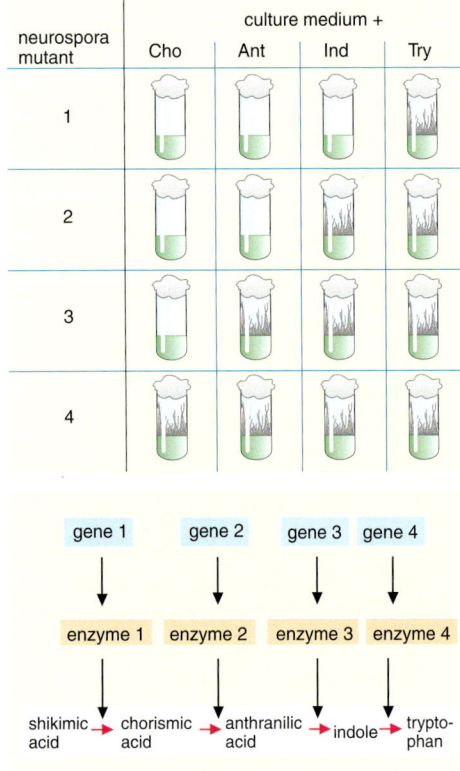

neurospora mutant	culture medium +			
	Cho	Ant	Ind	Try
1				
2				
3				
4				

gene 1 → enzyme 1
gene 2 → enzyme 2
gene 3 → enzyme 3
gene 4 → enzyme 4

shikimic acid → chorismic acid → anthranilic acid → indole → tryptophan

1 Deficiency mutants of neurospora

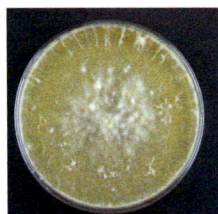

Neurospora

calcitonin
↑
mRNA [2 3 4]   parathyroid
↑
calcitonin gene
1 2 3 4 5 6
↓
mRNA [2 3 5 6]   nerve cells
↓
CGRP

2 One gene, two gene products by splicing

From genotype to phenotype

Genes determine distinct characteristics in an organism. All genes together form the *genotype* of an organism; its characteristics produce the *phenotype*. At first, most genes are transcribed into mRNA, which is then translated into polypeptides. However, only fully functional proteins control the expression of character states. In prokaryotes, almost the entire DNA is translated directly into RNA and then into polypeptides. Eukaryotes, in contrast, only use a small part of their DNA to code for proteins and often the translation process is not a direct one because different mechanisms of RNA editing can be inserted between gene and protein. The same DNA base sequence can control different characteristics. We therefore say that the genotype and the phenotype of eukaryotes are "soft-wired". Historically, various experiments have led to contradictory ideas about the gene and have made changes to the definition of the gene necessary.

Gene
A DNA segment coding for an RNA (see page 16)

Genotype
All genes (hereditary information) of an organism

Phenotype (appearance)
All characteristics of an organism

Proteome
Complete set of proteins expressed by a cell type

Genome
The complete genetic information of a cell including both coding and non-coding sequences of the DNA

One gene – one polypeptide hypothesis

In 1941, the geneticists GEORGE W. BEADLE and EDWARD L. TATUM found strains of the mould *Neurospora* that could not synthesize the essential amino acid tryptophan from shikimic acid. However, growth was induced after purposely adding intermediates of this pathway (fig. 1). The researchers concluded that these strains lack enzymes involved in this pathway and that this is the result of defects in the associated genes. The feature "successful tryptophan synthesis" is based on a chain of enzyme-coding genes. This is a *polygenic characteristic*, with each gene coding for one enzyme.

Splicing provides flexibility

The calcitonin gene is active in cells of the *parathyroid* and the nervous system (spinal cord). It consists of six exons (fig. 2). In the parathyroid, only exons 2 to 4 appear in the mature mRNA and are then translated into the hormone calcitonin. However, in nerve cells, the mature mRNA contains only exons 2, 3, 5, 6 — this is called *cell-type-specific splicing*. The product after translation here is the neuropeptide CGRP (*calcitonin gene related protein*). Based on one and the same gene, different gene products are thus produced in various tissues.

RNA editing

RNA editing changes the base sequence of a later transcribed mRNA. Two variants of apolipoprotein B with different functions are known: a larger variant (ApoB-100) expressed in many different mammalian cell types (binding partner of cholesterol) and a smaller variant (ApoB-48) only present in the cells of the small intestine and involved in lipid catabolism. Both variants are coded for by the same mRNA. In contrast to all other cells in the body, one cytosine of the mRNA is changed to uracil in cells of the small intestine. The respective sense codon is transformed into a stop codon (termination codon), and translation is thus prematurely terminated. A shorter protein molecule is produced.

The proteasome – specific protein degradation

Thousands of different proteins are present in a cell. Analogous to the term *genome*, the term *proteome* describes the totality of all proteins of a cell. Proteins can interact with each other inside a cell and can develop superior structures such as enzyme complexes. These are usually not examined by molecular biological methods but rather with biochemical methods (*electrophoresis, crystallography*) and also with electron and immunofluorescence microscopy.

The proteasomes found with these methods degrade unnecessary proteins in the cell (A. Ciechanover, A. Hershko, J. Rose; Nobel prize 2004). They are present in the nucleus and cytoplasm of virtually all types of organisms. Proteasomes are large cylindrical complexes composed of dozens of proteins. They have a canal in the middle. Proteins that should be degraded, e. g. unnecessary enzymes, are first marked in the cell by the addition of the protein ubiquitin. This complex is recognized and bound by the cap-shaped ends of the proteasome. To be degraded, proteins are transported into the canal and thereby are unfolded. Some of the proteins that form the canal are *proteases*, thus enzymes that cut proteins. The middle part represents the catalytically active component. The remaining amino acids can be reused by the cell to produce new proteins.

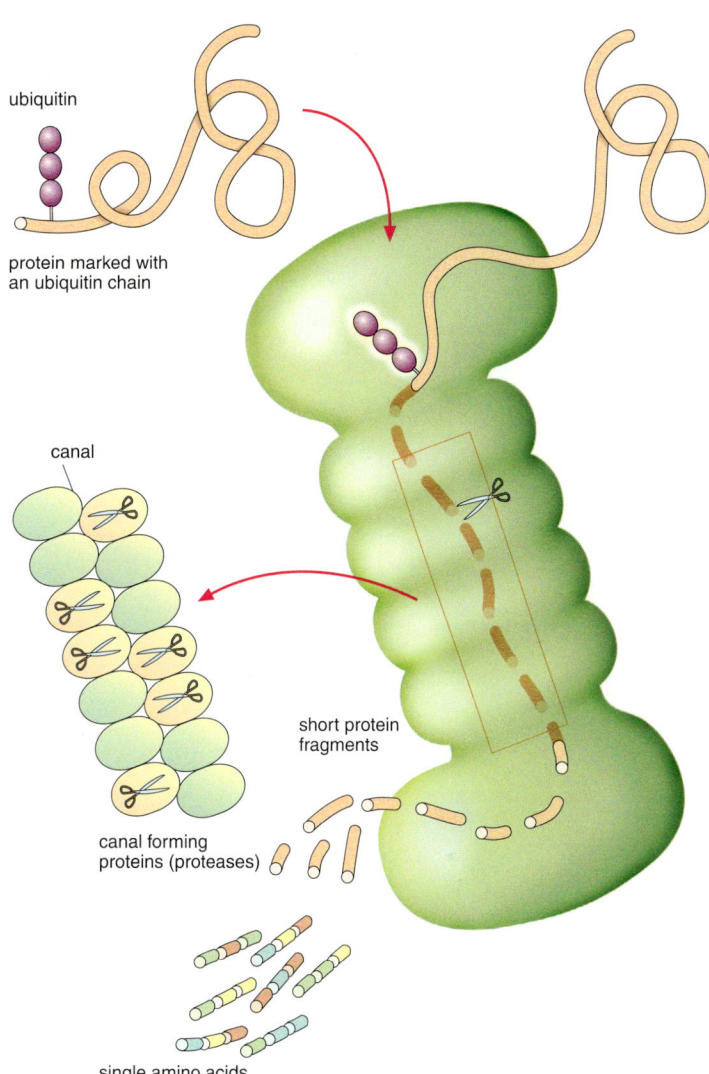

ubiquitin

protein marked with an ubiquitin chain

canal

short protein fragments

canal forming proteins (proteases)

single amino acids

1 Proteasomes degrade marked proteins in cell

Tasks

(1) Four enzymes are involved in the shikimic acid-tryptophan pathway (fig. 26.1). The examined Neurospora strains have characteristic defects in the participating genes. Indicate the strain and the mutated gene in each case.

(2) Single genes can affect several traits *(pleiotropy)*. Explain the phenotypic consequences of the defect that leads to cystic fibrosis in humans (see pp. 68 – 69).

(3) The original molecular biology definition of a gene as "construction manual for a polypeptide" has had to be extended. Explain why.

Polygenic inheritance
Several genes control one feature

Pleiotropy
Single genes influence multiple features

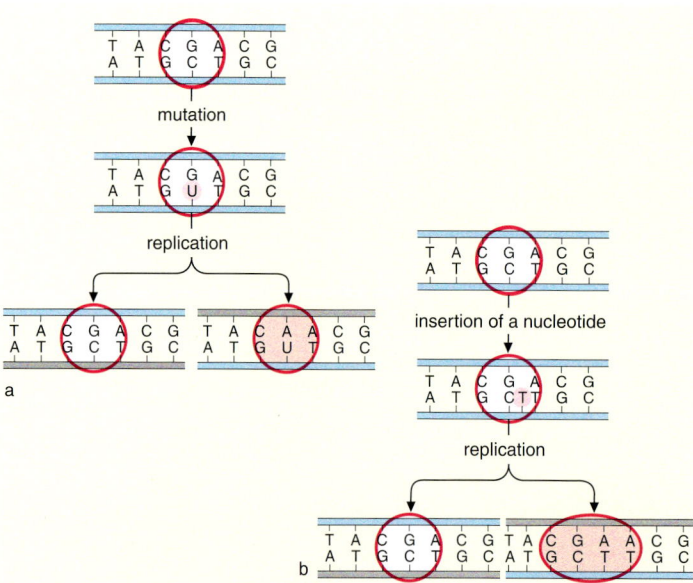

1 Point mutations (a) and frameshift mutations (b)

DNA damage and repair

Mutations are changes within the genetic information. They are caused by mutagens such as UV radiation and chemicals but also occur spontaneously because of the chemical instability of the DNA building blocks. Mutations occur *undirected*, it is not possible to predict where they will happen in the genome. Examinations of bacterial cultures have shown that, in one generation, 1-10 spontaneous mutations occur per 1,000,000 genes. Mutations can affect a single gene (*gene mutation*) or the structure

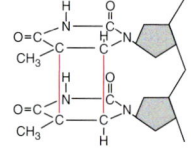

Thymine dimer

of a chromosome (*chromosomal mutation*) or the number of chromosomes (*genomic mutation*).

Gene mutations

Gene mutations only affect a single gene. *Point mutations* can be distinguished from *frameshift mutations*. Substitution of a single base is called a point mutation. In bacteria, about 70% of all mutations are point mutations and most deficiency mutants are attributable to such mutations. Based on the exchange of a DNA base, the amino acid sequence of the encoded enzyme is changed and therefore it often loses its function. Frameshift mutations include the addition (*insertion*) or removal of a base (*deletion*). In both cases, the reading frame of the mRNA triplet code is shifted if the change is not caused by three bases or a multiple of three. This results in a completely different protein. Whereas gene mutations in haploid bacteria directly affect the phenotype, diploid organisms are better protected: Most metabolic diseases are recessive, which means that the unchanged variant of the gene takes over the task of the mutated gene.

Task

① A point mutation does not always produce an effect, because of the redundancy of the genetic code. Such a mutation is called a *silent mutation*. Explain this using an example.

DNA repair mechanisms

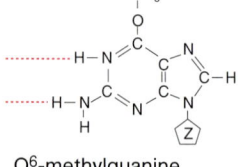

O6-methylguanine

Special DNA polymerases are able to cleave a wrongly incorporated non-complementary nucleotide out of the newly synthesized strand. This reaction takes place during replication if the base cannot form hydrogen bonds to the template strand.

UV light is absorbed by the purine- and pyrimidine-ring systems of the DNA bases. Adjacent thymines interconnect covalently to form dimers (see margin) that cannot form hydrogen bonds with adenine.

If the damage is not recognized, the next replication will stop at this point. The "repair enzyme" *photolyase* separates thymine dimers.

Of interest, this enzyme is activated by visible light. In nature, UV light never occurs without visible light.

The base guanine is changed to O6-methylguanine by *ethylmethanesulphonate* (EMS). O6-methylguanine pairs with thymine instead of cytosine. In the next replication, the base pair G-C is changed to A-T. In the bacterium *E. coli*, the enzyme called O6-methylguanine-DNA-transferase reverses the chemical change of guanine and thereby inhibits the point mutation. In the presence of EMS, the production of this *detoxification* enzyme is induced.

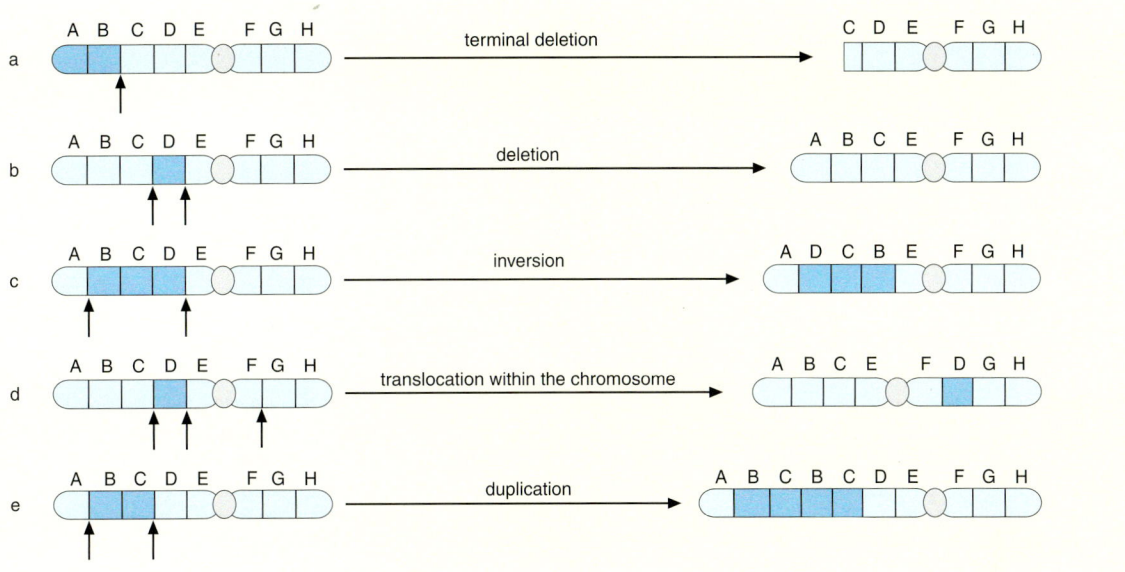

1 Chromosomal mutations (schematically)

Chromosomal mutation: modification by breakage and fusion

DNA molecules can break and are then fused again within the cell. Sometimes, large DNA segments are lost or are wrongly recombined. In *imbalanced* chromosomal mutations, the number of genes is altered, whereas in *balanced* chromosomal mutations the number of genes remains unchanged. A change in the position of a gene can severely influence its effect or can even cause the total loss of its function.

A single break separates a chromosome into two pieces. Only one piece contains the centromere, the attachment site of the spindle apparatus *(terminal deletion)*. The piece lacking the centromere is lost during the next cell replication cycle. The piece with the centromere can be transported into the daughter cell and be replicated. However, it lacks one terminal telomere and the open chromosome end "sticks" to the open chromosome end of the sister chromatid. The elongated fusion product now contains two centromeres and is easily ripped apart during the next mitosis.

If a chromosome breaks twice, an internal chromosome fragment can be lost *(deletion)*. The free ends then fuse. A severe disease in humans called *Cris-du-chat syndrome* is caused by a deletion in chromosome 5. If a fragment caused by two breaks is turned through 180° and then re-fused

into the same chromosome, we call this an *inversion*. The number of genes remains unchanged but the order of the genes is altered. During meiosis, recombination of the inverted part is more complicated and it is therefore often inherited as an unchanged unit ("*super gene*"). Chromosomal mutations are readily detectable in the banding pattern of giant chromosomes that form pairs of changed and unchanged homologous chromosomes, as observed by light microscopy (see margin).

Chromosomal fragments can also be re-fused in another position of the same chromosome or another chromosome *(translocation)*. In extreme cases, two whole chromosomes fuse together as seen in *translocation trisomy* (see page 66).

A special case is *duplication* in which a chromosomal fragment was doubled before and then re-fused into the chromosome. The multiplication of a nucleotide sequence causes e. g. *Huntington's chorea* (*St. Vitus' dance*, see page 78), a hereditary disease. Duplication can also lead to the production of new genes if a gene and its duplicate are changed by mutations. *Gene families* form such as the globin family, which codes for elements of mammalian haemoglobin.

Giant chromosome with chromosomal mutation (inversion)

wild einkorn wheat:
genome AA,
2n = 14 chromosomes,
only about 20 grains per ear,
brittle spikelet, grains strongly
attached in the ears.

wild grass 1
genome BB

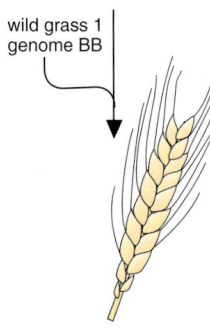

emmer: genome AABB,
4n = 28 chromosomes,
plenty of protein, disease
resistant, brittle ears,
precursor of Durum wheat
(firm ears and high harvest)

wild grass *2*
genome DD

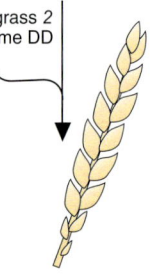

spelt: genome AABBDD,
6n = 42 chromosomes,
firm ears, grains difficult
to thresh out, grows in
different climates, good
for baking

gene mutations

bread wheat:
genome AABBDD,
6n = 42 chromosomes,
firm ears, easy to thresh,
very high harvest
**Evolution of
bread wheat**

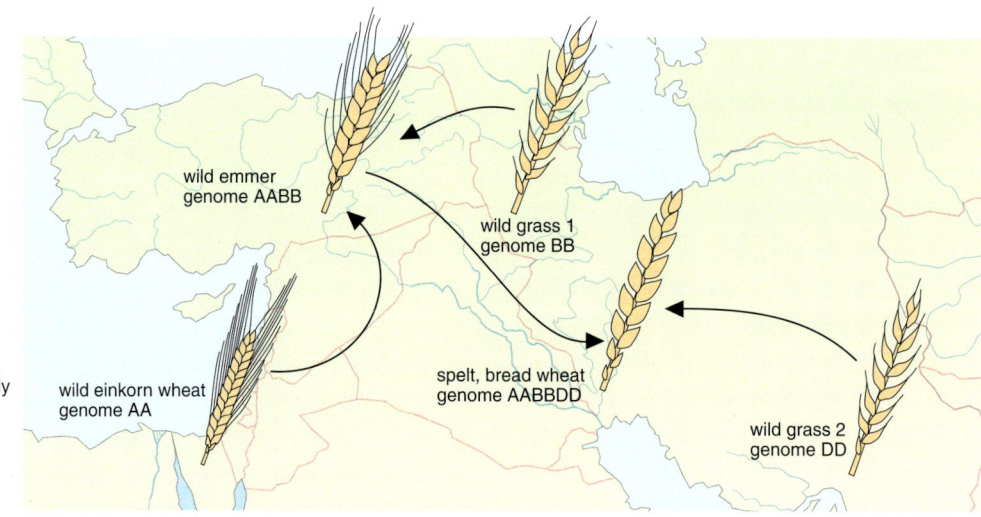

wild emmer
genome AABB

wild grass 1
genome BB

wild einkorn wheat
genome AA

spelt, bread wheat
genome AABBDD

wild grass 2
genome DD

1 Evolution of the bread wheat

Genomic mutation: change in the number of chromosomes

In a cell, the proportion of the quantities of DNA and cytoplasm allows an optimal regulation of the metabolism. This nucleus-cytoplasm relation finally determines the size of the cells. The amount of DNA in the nucleus of eukaryotes is set by the number and size of the chromosomes. A change in the number of chromosomes may be part of normal development but can also be based on mutations.

In most animals and higher plants, two sets of homologous chromosomes are present in somatic cells, which are therefore referred to as *diploid* (2n). Only gametes have a single set of chromosomes, i. e. they are *haploid* (1n). Mosses and ferns regularly alternate between generations of diploid and haploid cells (see page 35). The latter are often more delicately built. This is also true for plants with artificially introduced haploidy, e.g. by pollination with sterile pollen. Haploid mutants of animals, however, do not survive usually.

If cell division is not carried out after mitosis or meiosis, a multiplication of the set of chromosomes in the nucleus occurs (polyploidy). Cells, tissues and fruits of polyploid plants are particularly large. This is the reason that artificial polyploidization is a widely used method in plant breeding. Cell toxins such as the alkaloid colchicine derived from the autumn crocus (*Colchicum autumnale*) are used to inhibit the formation of the spindle apparatus and to obtain plants with large fruits. In humans, an egg cell is occa-

sionally fertilized by two sperm. The zygote then carries a triple set of chromosomes and therefore cannot survive (69, XXY or 69, XXX).

If chromosomes are falsely distributed to the daughter cells during cell division, a single chromosome might be added or lost. The lack of a chromosome usually has more dramatic effects than the addition of a third chromosome (*trisomy*). In tomatoes, trisomies of all chromosomes are known and the different kinds are phenotypically distinguishable. In humans, only a few trisomies of autosomes are able to survive but they result in severe deformities: Trisomy 21 (*Down syndrome*), Trisomy 18 (*Edward's syndrome)* or Trisomy 13 (*Patau syndrome*) (see page 66). In humans, an additional or a missing gonosome *(sex chromosome)*, however, is not so harmful. Examples are the *Klinefelter's syndrome* 47, XXY or *Turner syndrome* 45, X0. Several additional gonosomes rarely occur.

Tasks

1. The combination of sets of chromosomes of closely related species results in a non-homologous polyploidy. Explain the evolution of bread wheat by using figure 1.
2. Look up the symptoms of Klinefelter and Turner syndrome.

Mutagens

Experiments, for example, show that *mutagens* cause mutations in all organisms, mainly based on the same principle. DNA is the carrier of genes and, as a macromolecule, always shows the same reactions. Mutations in somatic cells can damage the cell so severely that it dies. Cell loss can be compensated by the division of adjacent cells. However, if genes controlling cell growth are affected, tumours can form. Mutations in gametes have no consequences for the affected organism but for its offspring.

Chemical mutagens

Base analogues are base-like substances that function as placeholder during replication. 5-Bromouracil (5BU) is derived from thymine whose methyl group is substituted by bromine. 5BU is integrated during replication instead of thymine. As can be expected, 5BU can pair with adenine but, after spontaneous chemical rearrangement, it can also pair with guanine. If this happens during replication, one of the DNA daughter molecules contains a 5BU — G instead of the original A — T base pair. After

5-Bromouracil (keto form)   adenine

5-Bromouracil (enol form)   guanine

a further replication, the daughter molecule carries a point mutation, the G — C pair, which is passed on in the following replications.

Ethidium bromide (EB) is a dye that makes DNA bands in an electrophoresis gel visible. The EB molecule has the size of a base pair and can move in between the DNA base pairs. This causes stretching and unwinding of the DNA double helix, which often leads to a DNA replication stop. Furthermore, additional nucleotides are often incorporated resulting in a frameshift mutation (see page 28). Electrophoresis gels stained with EB have to be disposed off as hazardous waste. *Benzanthracene*, a substance present in cigarette smoke, causes similar effects.

ethidium bromide

tar, benzanthracene

Physical mutagens

Heat is one of the most important mutagens. Heat breaks the bonds between the purine bases (A, G) and their sugars. DNA segments without purines cannot be repaired and can cause point mutations or deletions during replication. We estimate that, in a human cell, up to 10,000 segments with purine loss can evolve per day.

Radiation

On entering a cell, **X-rays**, **gamma radiation** and **corpuscular rays** (alpha and beta rays) release energy. If this energy meets a macromolecule such as DNA, it can have severe effects. Strand breaks, covalent linkage between nucleotides and the destruction of DNA bases are observed. Indirect damage can occur if the radiation hits water molecules in the cell. This releases hydroxyl radicals ($\cdot$OH) that cause oxidative DNA damage.

Radiation researchers have identified about 100 different reaction products of DNA nucleotides caused by ionizing radiation. One of these is 8-oxoguanine (8-oxoG). During replication, an 8-oxoG in the template strand can pair with a cytosine but also with an adenine nucleotide.

guanine   8-oxoguanine

Ames test: mutagenic potential

In 1970, B. N. AMES developed a simple test to assess *mutagenic potential*. He used a strain of the bacterium *Salmonella typhimurium*, which cannot synthesize the amino acid histidine because of a point mutation. The deficiency mutant cannot grow on culture media lacking histidine and therefore does not form colonies. If mutagenic substances are added to the culture medium, single *back mutations* can occur. The number of back mutations can be directly counted as the number of colonies on the agar plate. The higher the mutagenic potential of a tested substance, the higher is the number of developing colonies. When assessing chemicals, however, it is important to keep in mind that many substances only become mutagenic after being metabolized by the liver.

For example benzo[a]pyrene, a tar substance present in cigarette smoke, is modified by oxidation in the liver and can then move in between the base pairs of the DNA. This is the reason that tested chemicals are first pretreated with liver extract in order to estimate their mutagenic potential in the human body (in vivo).

Genes and the environment

Modification
Non-inheritable alteration of features attributable to environmental factors

Reaction norm
Genetic framework for feature expression

In 1871, HERMANN HENKING discovered that 50 % of the sperm of the firebug contained a special chromosome. He called it the X factor. Egg cells fertilized with these sperm developed into females. This was the first evidence for the connection between a chromosome and a trait. 90 years later, the change of the sex of female mice by transferring a single gene from male mice was possible. These transgenic mice were infertile but resembled males and also showed male behaviour. If genes decide on the sex of an organism, this is called *genotypic sex determination*.

inheritance (see page 38). In contrast, the flower colour of Petunias depends on environmental factors (*modification*).

Environmental factors can thus influence (modify) the phenotype of an organism. Modifications of a feature cannot be inherited. They are based on altered gene expression but not on changed genes. Today, we believe that genes contain the framework information for the possible expression of a trait that can be influenced differently by the interaction of genes and environmental factors. Genes decide the *reaction norm*. The presence of a specific gene may not lead to the same phenotype if the environment changes. Differentiation between modifications and inherited characters is often only possible if two genetically identical individuals (twins, clones) are studied under different environmental conditions. Some features have been established to be e*nvironmentally stable*, e. g. blood group in humans, whereas others are *environmentally unstable*, e. g. body weight.

The way that genes and the environment work together to allow the expression of features also poses an important question for society: is a person merely the product of its genes or a product of its environment by socialization and education? Many provoking questions that have been answered differently in various periods and cultures can be derived from this basis. Is it possible to avoid crime in a suitable environment? Is the preference of girls for social jobs based on one-sided gender-based education or on inborn caring behaviour? Is obesity inheritable or the result of adopted nutritional habits or both?

1 Flower colour of the Marvel of Peru and Petunia

(Figure labels: P, F₁, F₂ — 1 : 2 : 1; 30 days; 48 days; at 20 °C and full sun light; 64 days; continuously at 30 °C and weak light)

In contrast to this, observation of eggs of the European pond turtle (*Emys orbicularis*) shows that eggs incubated at 27.5 °C only result in males, whereas at temperatures over 29.5 °C, only females hatch. If the temperature is between 27.5 °C and 29 °C, both sexes hatch. The sex of many crocodiles and lizards also depends on the incubation temperature. In *phenotypic sex determination*, environmental factors decide on the sex of an organism.

As in the feature "sex", the environment and genes also affect other traits in different ways. For example, the Marvel of Peru blooms in red, white and pink. The character flower colour is an inherited feature following the Mendelian rules of intermediate

Task

① Explain, by using figure 1, what influences the flower colour of Petunias.

Petunia

Sex determination

Most seed plants and also many animals (earthworm, Roman snail) are *hermaphrodites*, i.e. the same individual produces female and male gametes. *Dioecious living beings* have separate sexes, and female organisms can be distinguished from males. In the beginning of their development, their cells are able to form both sexes. This bisexual potential is shifted in favour of the female or male sex depending on genes or environmental factors. If this process is interrupted, individuals can develop that are neither clearly male nor female (*intersexes*).

Genotypic mechanisms

XX/XY system

Males and females of many diploid animal species have two sets of *autosomes* and two *gonosomes* in their somatic cells. One of the two sexes has two different sex chromosomes and can therefore produce two different types of gametes. This is called *heterogametic sex* (Greek: *heteros* = different, Greek: *gametes* = germ cell). The sex that only produces one type of gamete is called *homogametic* (Greek: *homoios* = same).

lion XY, lioness XX

In all mammals, including humans, the female sex is homogametic (XX) and the male sex is heterogametic (XY). This is also true for many kinds of fish, beetle and grass frog. The males can produce sperm with different karyotypes: X or Y. In mammals, the presence of the Y chromosome causes the development of male sex organs. Only a specific part of the Y chromosome is responsible for this development. In humans, this region is called SRY and, in mice, Sry. SRY and Sry partly have corresponding base sequences but SRY does not work in mice. In the embryo, the expression of SRY/Sry causes the development of testicles from the early form of gonad (gonadal anlage). If SRY is missing, ovaries develop. The gonads produce hormones that control further sex differentiation.

ZW/ZZ system

Female birds are heterogametic (ZW). They can produce two different types of oocytes (Z or W). The homogametic males produce only one type of sperm (Z). This is e.g. also true for many reptiles, newts, butterflies and the African clawed frog.

peacock ZZ, peahen ZW

Haplodiploid system

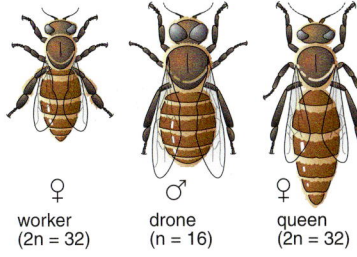

worker
(2n = 32)

drone
(n = 16)

queen
(2n = 32)

In bees and wasps, diploid females develop from fertilized egg cells. Unfertilized egg cells become haploid males. A sexually mature queen or a worker with rudimentary sex organs can emerge from a fertilized egg cell. Hereby, nutrition is important. Queen larvae are solely fed with food juice (royal jelly) produced by glands in the head of worker bees. The larvae developing into workers are additionally fed with pollen. The environmental factor 'nutrition' determines the different development of the females but not their sex.

Phenotypic mechanisms

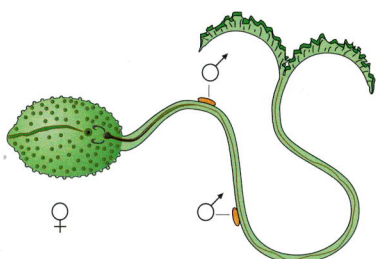

In *Bonellia*, a marine worm, sex determination depends on environmental factors. In open water, the larvae develop into females of about 15 cm in length equipped with a proboscis (ending in two flaps) that can be extended to about 1 meter for feeding. If a larva attaches to the surface of a mature female worm, however, it develops into a male of only 1 to 3 mm in size. Chemical signals on the body surface of the mature female are responsible for this process.

Sex change is well known in clown fish (see picture), which live in coral reefs. They are actually hermaphrodites. In an externally apparent male clown fish, the production of oocytes is inhibited by the presence of a larger female in the same territory. If the female is removed, the male changes into a spawning female.

Asexual and sexual reproduction

Reproduction
The process of gener-ating offspring and the transfer of genetic information from one generation to the next

Life only emerges from life and is based on the transfer of genetic information from generation to generation. Like growth and metabolism, *reproduction* is an essential characteristic of life. Genetic information has to be passed on reliably to the offspring so that rabbits produce rabbits, and anemones produce anemones. Although genetic continuity is essential for reproduc-tion, variability is also important. Whether features grant improved reproductive chances to the offspring mostly depends on environmental factors.

Asexual reproduction

In *asexual* (*vegetative*) *reproduction*, the offspring arises from only one parent and contains a copy of the parental genetic predisposition. Bacteria divide into daughter cells. Eukaryotic unicellular organisms evolve from the mitotic division of a parental cell into two daughter cells. Asexual reproduction is always connected to an increase in cell number. DNA is doubled prior to every cell division. In asexual reproduction, genetic variability is based on new mutations; this is rarely advantageous. This kind of reproduction guarantees continuity of genetic information. Many multi-cellular organisms also reproduce asexually: fresh-water *Hydra* can produce buds on the body wall; these buds detach after just a few days and develop into

Polyp stage of the moon jellyfish

independent animals. The buds emerge from hardly specialized cells. Such undiffe-rentiated cells that are capable of dividing are called *stem cells*. The use of biotechni-cal methods allows complete, genetically identical organisms to be obtained from stem cells of, for example, vertebrates (*clones*). Asexual reproduction is very common in plants. It is used in gardening by dividing up rootstocks and by taking cuttings, stolons, runners or rhizomes in order to obtain identical plant material. Multi-cellular plants and fungi can even emerge from a single cell that develops without prior fertilization. Such cells are referred to as *spores* and are produced in reproductive organs called *sporangia*. Spores are adapted for dispersion and can survive in unfavourable conditions for extended periods of time:

Sexual reproduction

Cells that are only able to develop into a complete organism after fertilization are called *gametes*. They are produced in specialized organs that are termed *game-tangia* in plants and fungi and *gonads* in animals. The process of fertilization is typical for *sexual reproduction*. The genomes of two parental gametes merge during fertilization and a zygote forms with newly combined genetic information. Sexual reproduction increases the variability of features in a population. Prior to every gamete production, the stock of chromo-somes is halved to a single set. Without this process, the genome of the offspring would always double after fertilization. Chromo-some reduction takes place during a special type of cell division called *meiosis* (see page 37). Sexual reproduction appears in unicellular and multi-cellular organisms. In many algae, all gametes look the same. Usually there are two types of gametes: a female gamete (*oocyte, egg cell*), which is large and immobile and a male gamete (*spermatocyte, sperm cell*), which is smaller and mobile.

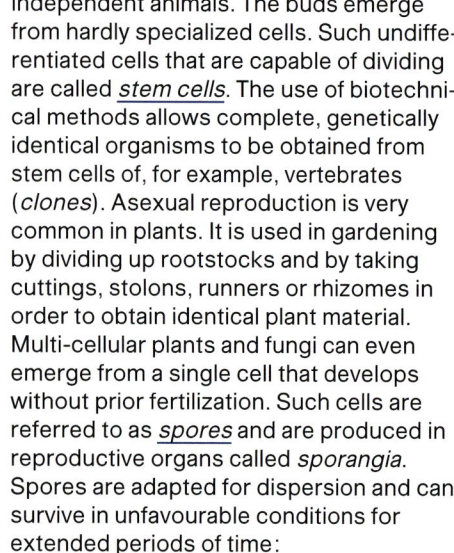

1 Medusa stage of the moon jellyfish (*Aurelia*)

Spores

Reproductive cells that develop into new individuals without prior fertilization. Spores can be haploid or diploid

Gametes

Reproductive cells that develop into new individuals only after fertilization. Gametes are haploid

Alternation of generations

Some unicellular algae (*Chlamydomonas*) usually reproduce asexually. However, if environmental conditions worsen, they switch to sexual reproduction and produce new variants. The probability of better adapted individuals is thus increased. The switch from asexual to sexual reproduction is not always caused by environmental factors. In numerous species, it is a fixed part of their development and is genetically predetermined. In sea hydrozoans (phylum: Cnidarians), polyps reproduce asexually by crosswise budding off. The plate-like daughter individuals become piled up on each other. After detaching, they become free-swimming jellyfish (*medusa*). The medusa have gonads in which they produce gametes for sexual reproduction. From each zygote, a new polyp emerges. Generations created by asexual and sexual reproduction alternate regularly (*alternation of generations*).

Alternation in nuclear phase

Alternation of generation also occurs in plants. This is not easily visible because the sexual stage can be inconspicuous. The typical *fern plant* has feathered leaves and represents the asexual stage called the diploid *sporophyte*. In sporangia on the underside of the leaf haploid spores develop by meiosis. Following mitotic division only and without fertilization, these meiospores become delicate plants called *prothallia* (sing.: *prothallium*). They represent the haploid *gametophyte*, the sexual stage. They produce haploid male and female gametes by mitotic division. After fertilization, the zygote becomes a diploid fern plant (fig. 1). In *Gymnosperms* (conifers), the tree represents the asexual stage, namely the diploid sporophyte. By meiosis, it produces two different kinds of meiospores. The *embryo sac cells* occur in the *ovule* of the *carpel* and the *pollen grains* are found in the *stamens*. In the ovule of the plant, the embryo sac cells become small female gametophytes and produce oocytes by mitosis. The pollen grains are dispersed and, on another carpel, develop into male gametophytes producing sperm cells by mitosis. The zygote develops into a diploid embryo after fertilization. The embryo with its sheath is called a *seed* and becomes a conifer.

The regular change between diploid and haploid generation is referred to as *alternation in nuclear phase*. *Haplophase* is started by meiosis and *diplophase* by fertilization.

fern

sporangium (spore case)

meiosis

sori (sing.: sorus), cluster of sporangia

fertilization

sperm cell   oocyte

gametangium

prothallium
meiospore

conifer

meiospores (n, pollen grains) are dispersed and develop into male gametophytes (n)

meiospores (n, embryo sac cells) remain in the ovule and develop into female gametophytes (n)

meiosis

stamen

meiosis

carpel with ovule

cone with stamens

cone with carpels

embryo of the sporophyte in the seed

seedling

conifer

fertilization

pollination: tiny male gametophyte produces sperm cells (n)

tiny female gametophyte produces oocytes (n)

1 Alternation of generations in ferns and seed plants

Fertilization and meiosis

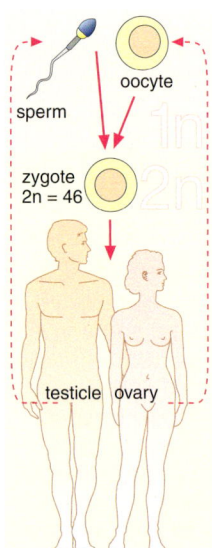

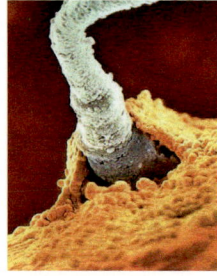

Human development

A characteristic of sexual reproduction is the fusion of two haploid gametes (n) forming a diploid zygote (2n). The fusion of the nuclei of the two gametes is called *fertilization*. Prior to fertilization (n + n → 2n), *insemination* occurs when a sperm cell penetrates the oocyte. Most animals and higher plants are diploids; only the gametes are haploid (see margin).

In order to reproduce sexually, the sexual partners have to recognize each other and must be synchronized in their readiness for mating. This requires complex patterns of sexual behaviour. Aquatic animals, such as fish and amphibians, release their gametes into the water and the sperms reach the oocytes by self-motion. This is called external insemination. Most terrestrial animals favour the direct transfer of sperm cells into the female reproductive organs in form of mating (*copulation*). During this internal insemination, the delicate gametes are protected from drying out and thus losses are reduced.

Gametes are produced in specialized organs *(gonads)* from diploid primordial germ cells. In order to produce haploid gametes, the two sets of chromosomes have to be reduced to a single set in an extraordinary cell division called *meiosis* (Greek: *meiosis* = reduction). As in mitosis, DNA duplication occurs prior to meiosis. At the beginning of meiosis, each primordial germ cell contains two sets of chromosomes (2n). The chromosomes consist of two chromatids (4C) respectively.

The chromatids are distributed randomly in the two separation steps (meiosis I, meiosis II) to the four gametes (1n, 1C):

— **Meiosis I:** the homologous chromosomes are distributed to two haploid daughter cells (1n, 2C) and the genetic information is newly combined.
— **Meiosis II:** the sister chromatids are separated by their centromeres and distributed to up to 4 haploid gametes (1n, 1C).

Male and female gametes then differentiate differently. The oocyte with various helping cells and sheaths forms the *egg*. The spermatocytes develop into flagellated *sperms*.

Tasks

① State the similarities and differences between the development of female and male gametes (page 37).
② Make a table and compare the processes during mitosis and meiosis.

»info box«

Fertilization in mammals

During insemination, the sperm attach, with the help of receptor molecules, to surface molecules of the outer coating that surrounds the oocyte (*lock-and-key principle*).This ensures that only gametes of the same species can fuse. The adhesion activates the *acrosome reaction*: The front tip of the sperm's head called the acrosome penetrates the coating of the oocyte. As soon as the membranes touch, they fuse and the sperm head and mid-piece are taken up into the oocyte. Now the cortical reaction takes place: the oocyte produces a substance that makes the membrane impermeable for other sperms. This inhibits multiple fertilizations. The nucleus of the oocyte finishes meiosis II (1n, 1C). The oocyte and sperm nucleus replicate (1n, 2C) and fuse (2n, 4C). The zygote thus contains a maternal and a paternal set of chromosomes, whereas its cytoplasm and mitochondria are derived from the oocyte. Mitochondria sporadically derived from the sperm are usually degraded.

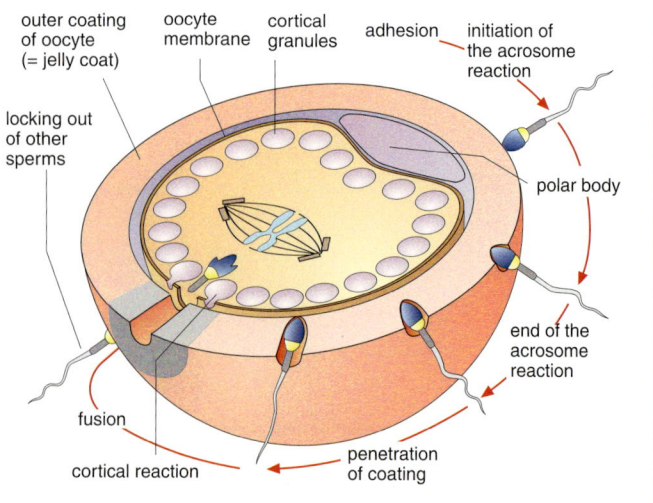

Meiosis: formation of haploid cells

Meiosis is used to reduce the sets of chromosomes of a diploid organism (2n) to 1n and to recombine the genes. It consists of two separation steps called meiosis I and meiosis II, which are again subdivided into several phases.

The results of meiosis are haploid gametes that can fuse and form a diploid zygote. The zygote grows by mitotic division and becomes an organism.

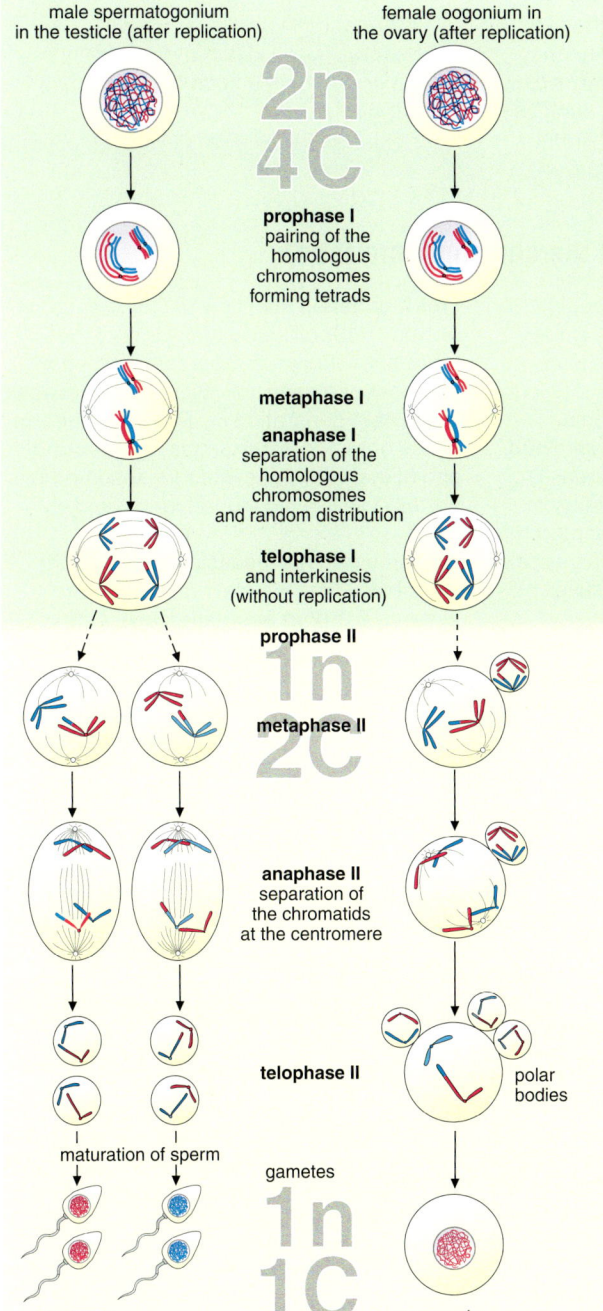

male spermatogonium in the testicle (after replication)

female oogonium in the ovary (after replication)

2n 4C

prophase I
pairing of the homologous chromosomes forming tetrads

metaphase I

anaphase I
separation of the homologous chromosomes and random distribution

telophase I
and interkinesis (without replication)

prophase II

1n 2C

metaphase II

anaphase II
separation of the chromatids at the centromere

telophase II

polar bodies

maturation of sperm

gametes

1n 1C

4 sperms

one oocyte

Meiosis I

Prophase I: The chromosomes begin to condense. It can take weeks until they are visible in a light microscope. The homologous chromosomes, which consist of two sister chromatids each, pair accurately. The paired chromosomes form tetrads of 4 chromatids; n *tetrads* can be counted by light microscopy. Each tetrad is held together by a complex of protein fibres that enables the exchange of DNA pieces between homologous chromatids. This so-called _crossing over_ (see page 43) is part of the recombination process of the genetic information. Later in prophase, the homologous chromosomes slightly drift apart and the locations where crossing-over have taken place are visible as *chiasmata* (crossings). Usually at least one chiasma occurs per homologous chromosome pair. The complex of protein fibres is degraded. The nuclear membrane and the nucleolus disintegrate at the end of prophase I and the *spindle apparatus* forms.

Metaphase I: The n chromosome pairs that recombined by crossing-over are arranged at the metaphase plate. This process is mediated by spindle fibres.

Anaphase I: The chromosome pairs are separated and then each set is transported to opposite cell poles by the spindle apparatus. Hereby, the partners are distributed to the poles randomly. The genetic information is thus recombined a second time.

Telophase I and interkinesis: At the cell poles, new nuclear membranes form around the chromosomes made of two chromatids (1n, 2C). The chromosomes slowly uncoil once again but do not replicate. They are therefore still microscopically visible in the daughter nuclei.

Meiosis II

Prophase II: The chromosomes, which are made up of two chromatids each, condense again and the nuclear membrane disintegrates.

Metaphase II: The chromosomes are arranged at the metaphase plate with the help of the spindle apparatus.

Anaphase II: The sister chromatids are separated at the centromere and pulled towards the opposite cell poles.

Telophase II: New nuclear membranes form around the chromatids. The nuclei of the gametes (1n, 1C) form.

character	opposed attributes (traits)
stem length	long : short
location of flowers	axial : terminal
pod shape	full : narrowed
pod colour	yellow : green
flower colour	violet : white

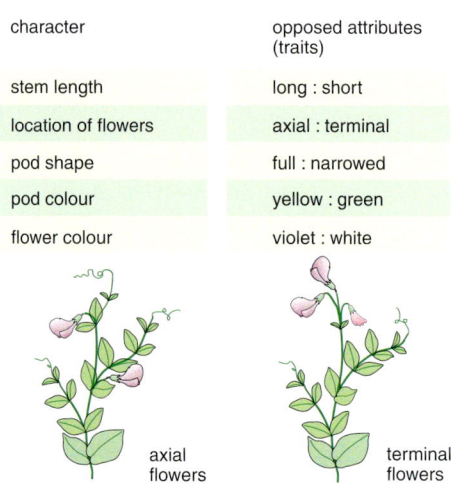

axial flowers   terminal flowers

JOHANN GREGOR MENDEL

1 Examined characteristics of the garden pea

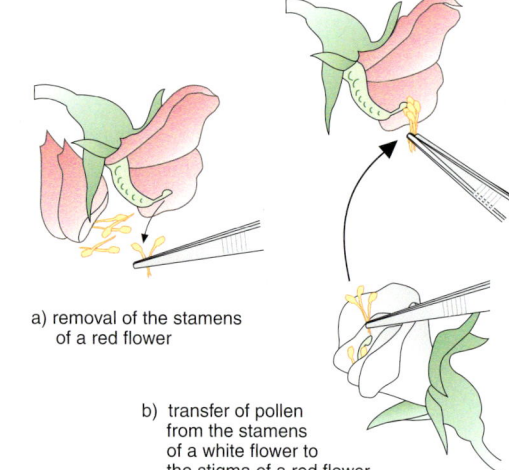

a) removal of the stamens of a red flower

b) transfer of pollen from the stamens of a white flower to the stigma of a red flower

2 MENDEL's method of cross-pollination

Mendelian inheritance and the chromosome theory

wrinkly and round peas

Allele
Alternative form of a gene

In the middle of the 19th century, people believed that the "cytoplasmic mix" is the reason that children have attributes from both mother and father. They were unable to imagine what these mixes were or that inheritance could be scientifically examined. The Augustinian monk JOHANN GREGOR MENDEL (1822 — 1884) in the monastery of Brünn had a different opinion. He wanted to know the rules that govern the appearance of parental features in offspring. MENDEL used pea plants (*Pisum sativum*) in his experiments. He chose varieties that have two clearly distinctive traits, for example, round/smooth versus wrinkly peas, that do not appear in a mixed (*intermediate*) form. Since MENDEL knew that he could expect hundreds of pea offspring, he chose features that could easily be evaluated. The features shown in fig. 1 are also stable, which means if the plant is self-fertilized, all offspring look like the parental plant. Because of self-breeding, they were *homozygous* (true-breeding).

Monohybrid cross

MENDEL used artificial crossbreeding to examine the inheritance of pairs of characteristics by, for example, taking pollen from plants with round peas and putting it onto the carpels of plants with wrinkled peas *(monohybrid cross)*. MENDEL's seven monohybrid crosses showed a clear pattern. The first plant generation (F_1) exhibited only one parental characteristic, e.g. all peas were round. They were uniform

(**1st Mendelian law**, *Law of Dominance*). If MENDEL self-pollinated the plants of the F_1 generation, however, characteristics that had not been seen in F_1 were apparent again in the next generation F_2. From the uniform F_1, he obtained, for example, 5,474 round and 850 wrinkly peas. Mendel observed this pattern in all seven monohybrid crosses.

He compared his results and came to the conclusion that the real ratio was approximately 3:1 (**2nd Mendelian law**, *Law of Segregation*). From this consistent pattern of inheritance, MENDEL concluded that each feature must be controlled by internal factors that occur in pairs. These factors can appear in various forms (e. g. R for round and r for wrinkly peas). Since one parental characteristic disappeared in the outward form of F_1 and reappeared in F_2 he concluded that the plants in F_1 must have two forms (R and r).

The Mendelian factors are referred to as *genes* in molecular biology. Most eukaryotic cells are *diploid*, which means that they have two homologous copies of each chromosome. Genes of eukaryotic organisms can exist in two variants called *alleles*. The genotype (Rr) is referred to as *heterozygous*, and (RR) and (rr) as *homozygous*. The genotype (Rr) expresses the visible feature (phenotype) "round peas". The allele (R) is dominant over the allele (r): alleles can be dominant or recessive. If there are phenotypic mixed forms, the alleles are called *intermediate*.

organism: pea

1st character: pea colour

alleles:
A: allele for the trait yellow pea colour

a: allele for the trait green pea colour

2nd character: pea shape

alleles:
B: allele for the trait round pea

b: allele for the trait wrinkly pea

pattern of inheritance: both dominant-recessive and freely combinable crossing scheme regarding the 3rd Mendelian law

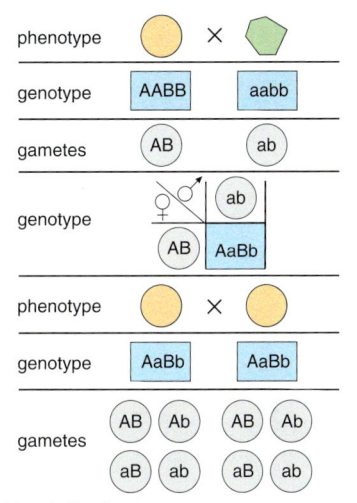

P

phenotype ○ × ⬠

genotype | AABB | aabb

gametes (AB) (ab)

genotype

	♂ (ab)
♀	
(AB)	AaBb

F₁

phenotype ○ × ○

genotype | AaBb | AaBb

gametes (AB) (Ab) (AB) (Ab) / (aB) (ab) (aB) (ab)

distribution of genotypes

F₂

♀ \ ♂	(AB)	(Ab)	(aB)	(ab)
(AB)	AABB	AABb	AaBB	AaBb
(Ab)	AABb	AAbb	AaBb	Aabb
(aB)	AaBB	AaBb	aaBB	aaBb
(ab)	AaBb	Aabb	aaBb	aabb

phenotypes ○ ⬠ ○ ⬠

9 : 3 : 3 : 1

1 Crossing scheme regarding the 3rd Mendelian law

MENDEL looked for an explanation for the 3 : 1 ratio in the F₂ generation. The plants of the F₁ generation have the genotype (Rr) and a F₁ crossing can be described as follows: (Rr) x (Rr). From this crossing, peas can occur that have the genotype (rr). They have inherited a recessive allele from both parents. We know today that the alleles of the parents are separated during the production of the gametes (*meiosis*). During fertilization, male and female gametes fuse forming a zygote. Thus, the two parental alleles are brought together.

Tasks

(1) MENDEL also carried out experiments with pea plants that differed in two features (*dihybrids*). Derive from figure 1 the **3rd Mendelian law** (*Law of Independent Assortment*): Feature pairs can appear in new combinations in a dihybrid cross.

(2) There are many more genes than chromosomes. This means that numerous genes must be located on one chromosome and that they are then obviously inherited together (genetic linkage).
What changes would you expect in the given heredity if the allele pairs for yellow-green and round-wrinkly were located on the same pea chromosome?

»info box«

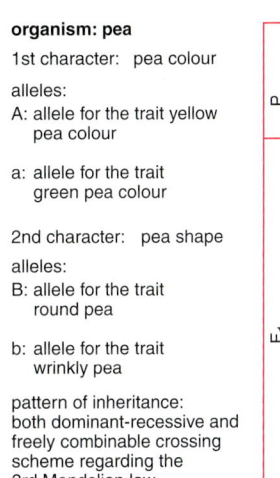

gene

F

wild type   mutant

Interpretation of dominance by means of physiology and molecular biology

MENDEL postulated that the characteristics of his round versus wrinkly peas were based on hypothetical factors of inheritance. Today we can trace these differences back to the gene product and the gene.

In round peas, saccharose is changed to starch via an enzyme. In wrinkly peas, this enzyme is not active because of a mutation in its gene, and the saccharose concentration always stays the same. For this reason, the peas take up more water by osmosis during growth. When they dry out, they lose the water again and shrink. The (r) allele is the mutated form. A plant of the genotype (rr) does not convert saccharose into starch. Heterozygous (Rr) plants still have

a sufficient enzyme concentration to lower the saccharose concentration. (Rr) plants have round peas; hence, (R) is dominant over (r). Inserting a DNA fragment (F) of the length of about 800 base pairs into the coding gene (see fig.) causes the structure of the starch-forming enzyme to change so substantially that it is no longer active.

Today we know a number of cases in which DNA fragments change genes in the described way and cause the loss of their function. Examples are spontaneous mutants of *Escherichia coli* and *Drosophila*.

Mendelian laws and statistics

Maize is an important plant in genetic research. Its unisexual flowers can be covered by paper bags in order to avoid uncontrolled pollination in experiments. The maize kernels contain the embryos of the successive generation. Kernel features are very suitable for inheritance experiments because they can be evaluated without growing the plant.

As in peas, smooth and wrinkly kernels are produced. The allele for smooth is dominant (allele A). Accordingly, the allele (a) for wrinkly is recessive. If heterozygous smooth F_1 plants are cross-bred, then the seeds of the F_2 generation must contain three parts smooth and one part wrinkly seeds according to the 2nd Mendelian law. The counting of three maize cobs resulted in the following:

	smooth	wrinkly	overall	ratio smooth : wrinkly	relative frequency smooth	relative frequency wrinkly
1. cob	209	76	285	2.75 : 1	0.733	0.267
2. cob	164	71	235	2.31 : 1	0.698	0.302
3. cob	201	56	257	3.59 : 1	0.782	0.218

Although the actual ratio 3 : 1 is never found, values come very close to it. The observed result depends on chance. Crosses are evaluated by counting the frequency of the studied feature. From this, the relative frequency as a ratio of smooth and wrinkly maize kernels to the overall count can be determined. The results approach the expected probability values (0.75 smooth and 0.25 wrinkly) according to Mendel's laws.

Events based on the chance of a new combination of alleles arising can be shown in computer simulations (fig. 1). Each point represents the relative frequency of smooth maize kernels in one set of crosses.

Result: the greater the number of offspring, the closer is the relative frequency to the probability value. Mendel analysed his breeding experiments already and examined many thousands of offspring.

Tasks

1. Calculate the relative frequency of wrinkly kernels by using the table for the maize cobs. Combine the results for cob 1 + 2, 1 + 3, 2 + 3, and 1 + 2 + 3. Are the relative frequencies comparable with the theoretically expected value?
2. A brown-eyed father and a blue-eyed mother are expecting a fourth child. They already have three brown-eyed children. They think that, according to the 2nd Mendelian law, their expected child has to be blue-eyed. What do you think about their idea?

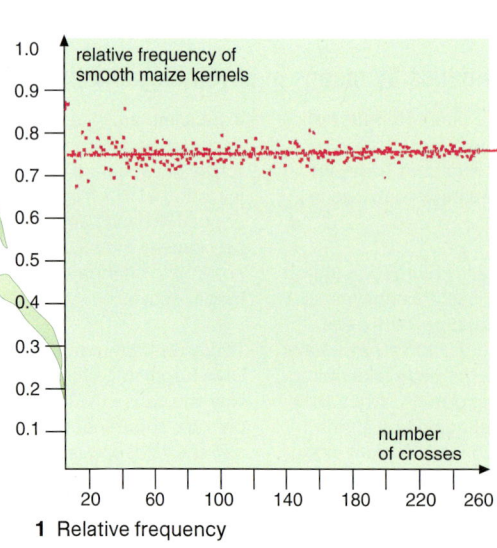

1 Relative frequency

2 Maize cobs

Exercise: Mendelian laws

Decide on one answer for each of the 10 following questions. If you look up the relative number, you will be able to see whether your answer is correct or incorrect. The number after the arrow shows where you have to continue.

1   A hobby gardener grows radishes of different shapes: long (genotype ll) and round (kk). The cross long x round results in oval radishes. What pattern of inheritance is this?
    dominant recessive → 13
    intermediate → 27

2   Since he prefers the oval radishes, he cross-breeds them with themselves in order to obtain only oval radishes. Will he be successful with this breeding method?
    yes → 23
    no → 20

3   How many (in percent) of the offspring are homozygous if, in three crosses, only the phenotypically identical F_2 radish plants are bred with each other?
    50 % → 19
    more than 50 % → 25

4   The gardener found out that the colour of the radish has an intermediate pattern of inheritance: red (rr), white (ww) and pink (rw). How many phenotypes are expected if oval pink-coloured radishes are cross-bred?
    4 phenotypes → 15
    9 phenotypes → 11

5   In what ratio will the 9 phenotype possibilities appear in the F_2 generation? Solve this question with a Punnett square.
    → 28

6   Round pink radishes were cross-bred. How many different phenotypes do you expect within the offspring of the next generation?
    3 phenotypes → 21
    6 phenotypes → 26

7   Which Mendelian law did you use for this cross?
    2nd law → 18
    3rd law → 16

8   From which cross will you only get pink and oval radishes?
    oval/pink x oval/pin → 12
    round/red x long/white → 24

9   Is it possible to obtain only oval and pink offspring from another cross?
    yes → 17
    no → 14

10 If you have successfully come this far, you should have no problems doing the Punnett square for trihybrid inheritance. What ratio do you expect for the phenotypes (A/B/C) with dominant and recessive alleles (Aa/Bb/Cc)?
    → 22

11 Correct! Each possible genotype belongs to a specific phenotype. *→ 5*

12 Incorrect! This cross would lead to further segregation (compare 28)). *→ 9*

13 Incorrect! Oval radishes are heterozygous (kl). If one allele was dominant, it would show itself in the phenotype. *→ 2*

14 Incorrect! The other homozygous combinations round/white x long/red also lead to uniform offspring. *→ 10*

15 Incorrect! There are 4 possible combinations in a dihybrid dominant - recessive inheritance pattern. There are 9 in intermediate inheritance: round/red, round/pink, round/white, oval/red, oval/pink, oval/white, long/red, long/pink, long/white. *→ 5*

16 Not necessarily wrong, because it is formally a dihybrid inheritance. However, it is better to use the 2nd law, because the phenotypes exhibit the ratio 1 : 2 : 1. *→ 8*

17 Correct! It is also possible to cross round/white x long/red. *→ 10*

18 Correct! The phenotypes exhibit the ratio 1 : 2 : 1. *→ 8*

19 Incorrect! The cross kl x kl leads, in 50 % of the cases, to homozygous offspring but ll x ll and kk x kk lead only to homozygous radishes. There will be more than 50 % homozygous offspring overall. *→ 4*

20 Correct! According to the 2nd Mendelian law, the phenotypes of the F_2 generation exhibit the ratio 1 : 2 : 1 within an intermediate inheritance pattern.
    → 3

21 Correct! Again, only round radishes with all three colours can be obtained. *→ 7*

22 Eight possible phenotype combinations (ABC : ABc : AbC : aBC : Abc : aBc : abC : abc) can occur in the ratio 27 : 9 : 9 : 9 : 3 : 3 : 3 : 1. *→ 29*

23 Incorrect! Oval radishes are heterozygous (kl). If they are crossed, the next generation (F_2) will exhibit round (kk) : oval (kl) : long (ll) = 1 : 2 : 1. *→ 3*

24 Correct! Both parents are homozygous. Their offspring are uniform.
    → 9

25 Correct! (Explanation see 19).
    → 4

26 Incorrect! There are only three combinations possible: round/red, round/pink, round/white. *→ 7*

27 Correct! The oval radishes are heterozygous (kl). This results in intermediate inheritance of an oval shape.
    → 2

28 The 9 phenotypes round/red, round/pink, round/white, oval/red, oval/pink, oval/white, long/red, long/pink, long/white occur in the ratio:
    1 : 2 : 1 : 2 : 4 : 2 : 1 : 2 : 1. *→ 6*

29 Thanks a lot for your patience!

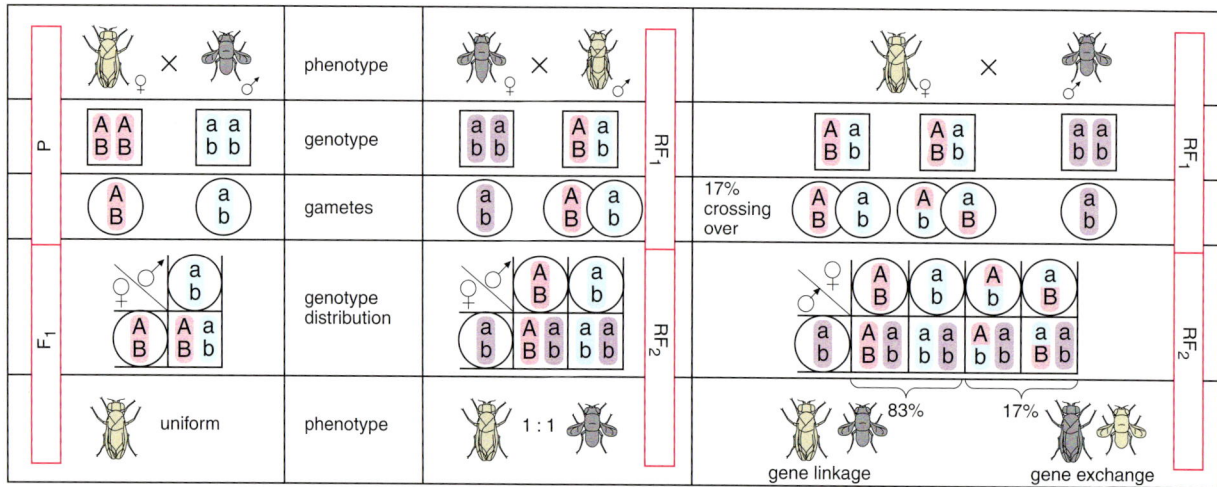

1 Linkage of genes in *Drosophila*

Linked inheritance and exchange of genes

The fruit fly *Drosophila melanogaster* is used as model organism in genetics. Basic mechanisms of inheritance were first examined in these flies. They are only a few millimetres in size, are commonly found on fruit and can be bred easily. The number of offspring is very high (500 eggs per female) and the generation interval is only 12 days. Apart from the *wild type* of the fly, there are many readily distinguishable mutants. Furthermore, *Drosophila* has only 8 chromosomes ($2n = 8$) in its diploid set and these can be easily dissected and examined.

The American biologist THOMAS HUNT MORGAN carried out thousands of experiments with *Drosophila*. When he crossed males with the two features, namely black body and vestigial wings, with true-breeding wild type females (normal body colour and normal wings), he obtained only wild type flies in the uniform F1 generation, as expected (1st Mendelian law).

The back cross of a male of the F1 and a female of the P generation resulted in wild type flies and black, vestigial winged flies in the ratio 1 : 1 (fig. 1). This result contradicts the 3rd Mendelian law, the law of independent assortment, stating that the genes are independently recombined. However, the result can be explained if we assume that several genes can be passed on by one chromosome. This is called a *linkage group* and the genes are written below each other in a crossing scheme (fig. 1).

MORGAN realized, from many crossing experiments, that *Drosophila* has 4 groups of linked genes. This fits with the number of chromosomes in the single chromosome set and confirms the chromosome theory of inheritance.

Further crossing experiments of MORGAN with black, vestigial winged mutants of Drosophila showed a surprising result. In the reversed back cross of a heterozygous female of the F1 and a male Drosophila mutant of the homozygous P generation, all 4 possible combinations of features appeared. The ratio was not according to MENDEL's theory (1 : 1 : 1 : 1) but most flies phenotypically resembled their parents. Only 17% of the offspring were single mutants (see fig. 1, right side). MORGAN first believed that he had made a mistake during the experiment but he always obtained the same result. Therefore the linked genes must have been separated *(breakage and reunion)*.

The chromosome theory provides a plausible explanation for these observations: during prophase I of meiosis, the homologous two-chromatid chromosomes pair so that DNA segments can be exchanged. This segment exchange is called <u>crossing over</u> and can be seen in metaphase I as a crossing structure (*chiasma*) when the homologous chromosomes separate. Alleles that used to lie on the same chromosome are distributed to different homologous chromosomes by crossing over. This happens in a controlled process in which enzymes

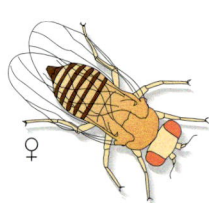

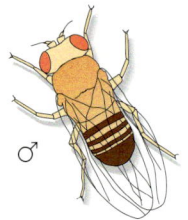

Drosophila (wild type)

make sure that the respective parts on the chromosome arms are broken and reunited anew (fig. 1). Thus, linked alleles can be inherited in a new combination.

This was confirmed when a special characteristic of the meiosis of two-winged flies, among them *Drosophila*, was discovered: chiasmata occur only during oogenesis and not during spermatogenesis. This is perfectly represented in the experimental results. Breakage and reunion occur only if the back cross is done with heterozygous females.

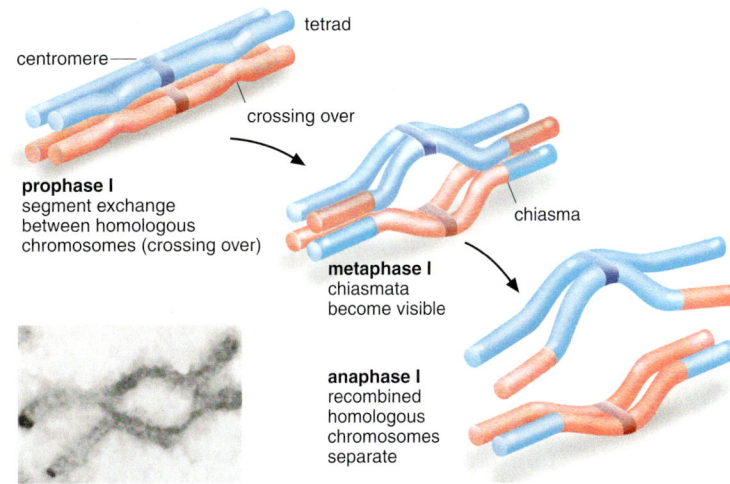

prophase I
segment exchange between homologous chromosomes (crossing over)

metaphase I
chiasmata become visible

anaphase I
recombined homologous chromosomes separate

1 Crossing over and subsequent chiasma formation

»info box«

Genetic map – the position of genes

When MORGAN counted the phenotypes, he found a different percentage of breakage and reunion for each gene pair. This characteristic value was called the *crossover value*. For example, the gene for light red eyes (cinnabar: cn) is linked with black (black: b) and vestigial (vestigial: vg) wings. The crossover value for b and cn is 9 % and for cn and vg is 9.5 %.

In order to explain the different crossover values, THOMAS HUNT MORGAN developed a theory. He assumed that the genes are located one after another in a line on a chromosome. Furthermore, he thought that the probability of crossing over is the same everywhere on a chromosome. Therefore the crossover value increases with the distance between two gene loci and can be seen as a relative measure of the distance between two genes on a chromosome.

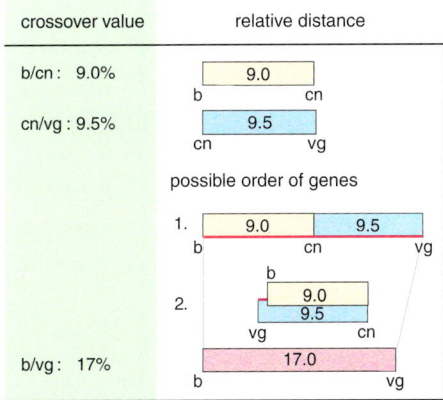

crossover value	relative distance
b/cn: 9.0%	9.0 (b — cn)
cn/vg: 9.5%	9.5 (cn — vg)
	possible order of genes
	1. 9.0 (b — cn) 9.5 (cn — vg)
	2. b / 9.0 / 9.5 (vg — cn)
b/vg: 17%	17.0 (b — vg)

→ the order of the genes is b-cn-vg

Using two crossover values, it is not possible to determine the order of three genes. Only a third breeding experiment, which, according to our example, shows the crossover value of vg and b, can solve the problem (*three-point analysis*). The result is 17 % and does not fit with the order vg-b-cn but fits well with b-cn-vg (see figure).

The crossover value b/vg (17 %), however, is smaller than the sum of the crossover values for b/cn and cn/vg (18.5 %). This value is expected when adding the relative distances. MORGAN assumed that more than one crossing over event can take place at the same time in a linkage group. Microscopic observations of chiasmata of the chromosome confirmed this. If one crossing over takes place between b and cn and an additional one occurs between cn and vg (*double crossing over*), the genes b and cn and the genes cn and cg are separated. However, the genes b and vg are still linked because they are located once again on the same chromosome. This is why the experimentally determined value is lower than the calculated one. If two gene loci are extremely far apart, breakage and reunion are as possible as linkage. Such genes do not seem linked.

Using three-point analyses, MORGAN and his team determined the relative distances of many gene loci and produced genetic maps of *Drosophila*'s four chromosomes. In the maps, the percentage crossover value of adjacent genes is given as dimensionless Morgan units (1 cM = 1 centi-Morgan).

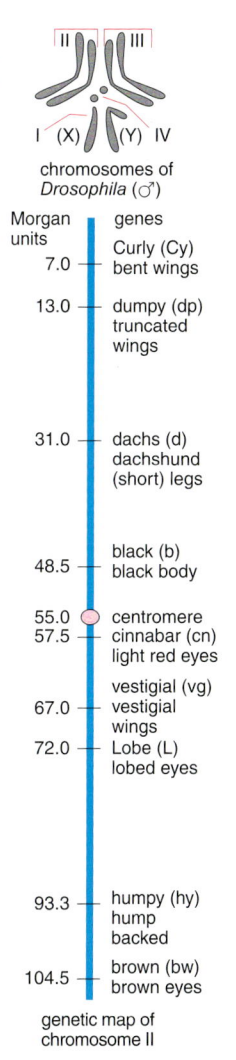

chromosomes of *Drosophila* (♂)

Morgan units	genes
7.0	Curly (Cy) bent wings
13.0	dumpy (dp) truncated wings
31.0	dachs (d) dachshund (short) legs
48.5	black (b) black body
55.0	centromere
57.5	cinnabar (cn) light red eyes
67.0	vestigial (vg) vestigial wings
72.0	Lobe (L) lobed eyes
93.3	humpy (hy) hump backed
104.5	brown (bw) brown eyes

genetic map of chromosome II

Drosophila genetics

Wild type and mutants

MORGAN introduced symbols in which the wild type is marked with "+". True-breeding wild type flies with brown body have two + alleles, whereas the ebony-coloured mutant has two recessive e alleles. In dihybrid inheritance, these symbols can lead to confusion because the two *different* wild type alleles carry the *same* description (+). If, for example, the ebony-coloured mutants are crossed with flies whose wings are curled, the allele pairs e and + show up in the crossing scheme, together with cu and +. In order to avoid confusion in such cases, some genetics publications use symbols that always state the wild type allele and its associated mutant allele: e and e^+ or cu and cu^+.

Body colour mutant "ebony"

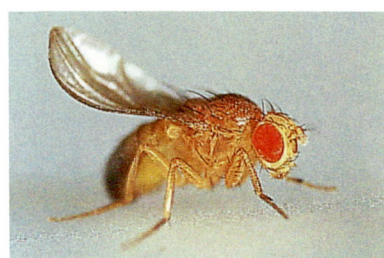

Wing mutant "Curly"

Tasks

① True-breeding ebony-coloured flies are crossed with true-breeding flies with curled wings. Find out, by using the genetic map, whether this is a case of gene linkage.

② Make a crossing scheme up to the F_1 generation. Wild type alleles are e^+ and cu^+ respectively.

Eye colours

Many mutants of *Drosophila* have been found with changed eye colour. Easily visible are the colours "cinnabar" (light red), "brown" and "white". Several wild type alleles are involved in the development of the wild type eye colour:
bw^+: allele for the development of a brown pigment
cn^+: allele for the development of a light red pigment
w^+: allele for pigment uptake into the pigment cells

Eye colour mutant "brown" (left) and "white" (right)

Wild type flies with red eyes have both pigments in their pigment cells of the compound eye. The allele w^+ is located on the X chromosome.

Tasks

③ Why do the flies with the genotype $\frac{cn}{cn}$ (all other alleles are wild type) have light red eyes?

④ True-breeding flies with light red eyes are crossed with true-breeding flies that have brown eyes. State the genotype and the eye colour of flies in the F_1 generation.

⑤ Males with the genotype $\frac{cn\ bw}{cn\ bw}$ are crossed with females of the F_1 generation. From this cross, the following phenotypes arise: wild type (281), brown eyes (281), light red eyes (293), colourless eyes (208). Explain these results by means of a Punnett square and calculate the crossover value of the genes cn and bw. Compare your result with the genetic map.

⑥ True-breeding females with the single mutation "white" are crossed with true-breeding wild type males. Find out, using a crossing scheme, which genotypes and phenotypes are present in the males and females of the F_1 generation. Take into consideration that male flies inherit their X chromosome from their mother and that there is no gene for eye colour on the Y chromosome.

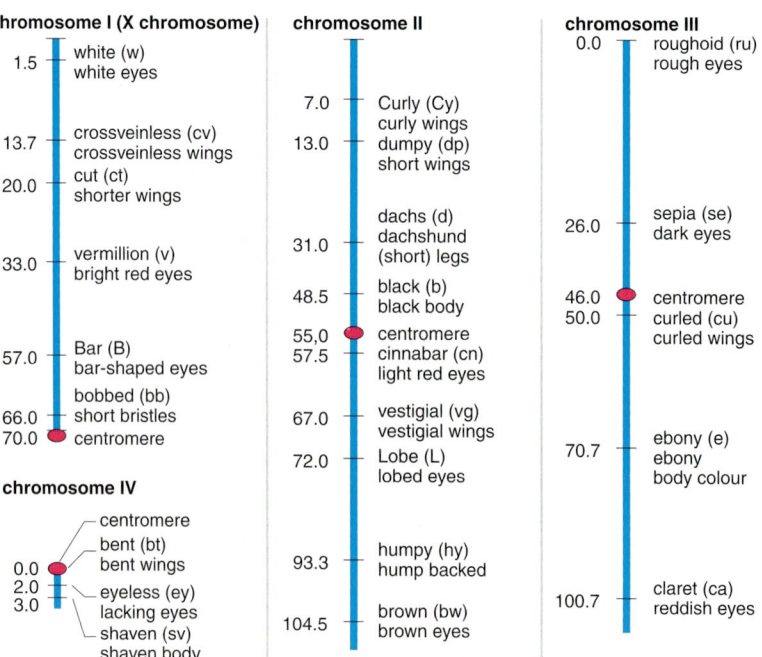

chromosome I (X chromosome)

1.5	white (w) white eyes
13.7	crossveinless (cv) crossveinless wings
20.0	cut (ct) shorter wings
33.0	vermillion (v) bright red eyes
57.0	Bar (B) bar-shaped eyes
	bobbed (bb)
66.0	short bristles
70.0	centromere

chromosome IV

0.0	centromere bent (bt) bent wings
2.0	eyeless (ey) lacking eyes
3.0	shaven (sv) shaven body

chromosome II

7.0	Curly (Cy) curly wings
13.0	dumpy (dp) short wings
31.0	dachs (d) dachshund (short) legs
48.5	black (b) black body
55.0	centromere
57.5	cinnabar (cn) light red eyes
67.0	vestigial (vg) vestigial wings
72.0	Lobe (L) lobed eyes
93.3	humpy (hy) hump backed
104.5	brown (bw) brown eyes

chromosome III

0.0	roughoid (ru) rough eyes
26.0	sepia (se) dark eyes
46.0	centromere
50.0	curled (cu) curled wings
70.7	ebony (e) ebony body colour
100.7	claret (ca) reddish eyes

Genes outside the nucleus

Following research in molecular biology, we know today that not all genes of eukaryotic cells are located in the nucleus. Chloroplasts and mitochondria have their own genes, the ability to replicate their DNA, the complete biosynthesis apparatus and the ability to reproduce by division. A number of plant species (petunia, tobacco, barley, marvel of Peru/*Mirabilis jalapa*) pass chloroplasts only via the egg cell to the next generation. During pollination, only the nucleus of the sperm from the pollen is handed on. This is called the *maternal inheritance* of chloroplasts and mitochondria. Rhododendron, however, passes chloroplasts on in both pollen and egg cells.

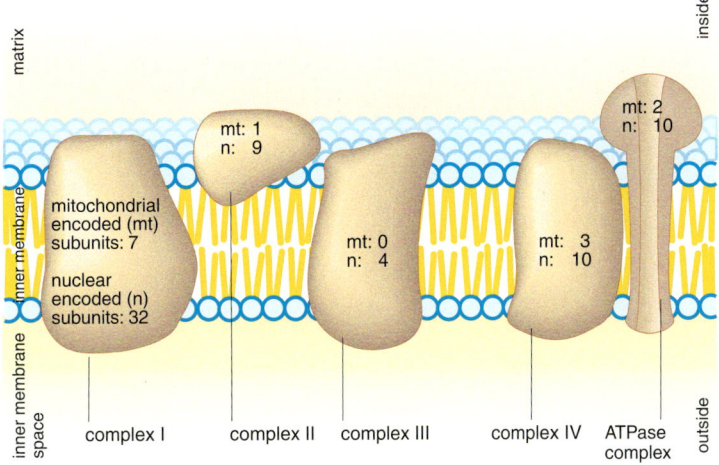

1 Origin of some enzymes of the respiratory chain

Genes in chloroplasts

In 1909, CARL CORRENS first observed maternal inheritance in *Mirabilis jalapa*. This plant species has green, white and green-white mottled leaves and branches. By microscopy, CORRENS saw that the green plant parts were made from cells with green chloroplasts and the white parts from cells with chloroplasts lacking chlorophyll (see margin). CORRENS pollinated carpels in flowers of green branches with pollen from flowers of differently coloured branches. He always obtained only absolutely green hybrids. If he pollinated carpels in flowers with white leaves, the offspring had only white leaves, no matter which pollen he used. If flowers of branches with green-white mottled leaves are pollinated, plants with green, white and green-white mottled leaves arise, no matter which pollen is used. CORRENS observed that the daughter plants always show the phenotype of the mother plant. Therefore he assumed that the gene responsible for chlorophyll production is not located in the nucleus but in the chloroplasts.

Genes in mitochondria

The sperms of most animal species contain several hundred mitochondria, whereas the egg cells possess several hundreds of thousands of mitochondria. Even if some mitochondria penetrate the zygote during fertilization, the mitochondrial genome of the offspring is almost completely derived from the maternal ancestors.

Cooperation with the nucleus

Chloroplast DNA contains about 100 protein-coding genes. However, the formation and function of the chloroplast require at least 1000 different proteins. Molecular biologists have found out that most of these proteins are encoded in the nucleus, produced in the cytoplasm and then transported inside the chloroplast. A typical example is the enzyme *Ribulose-1,5-bisphosphate carboxylase-oxygenase* (*RuBisCO*), which has a key function in carbon fixation during photosynthesis. The enzyme contains 8 large identical and 8 small identical subunits. The gene for the large subunit is located in the chloroplast genome and the one for the small subunit is in the nucleus. Similar cases occur in mitochondria. In humans, most of the mitochondrial proteins are also encoded in the nucleus. The enzymes of the respiratory chain are a mosaic of nuclear and mitochondrial encoded proteins (see fig. 1).

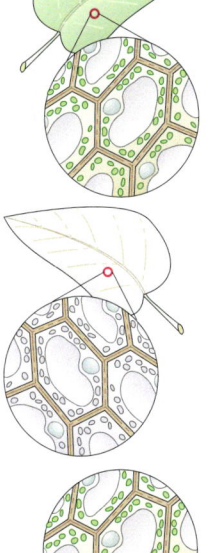

Tasks

1. The genetic apparatus of chloroplasts and mitochondria have some striking similarities and differences from the genetic apparatus of the nucleus. Use the information given on page 22 and make a table.
2. Maternal inheritance does not obey the Mendelian laws. Explain.

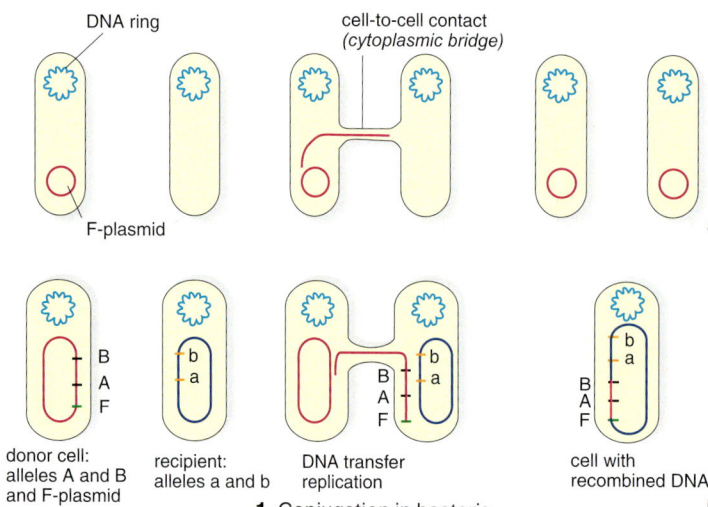

DNA ring

cell-to-cell contact
(cytoplasmic bridge)

F-plasmid

a

donor cell:
alleles A and B
and F-plasmid

recipient:
alleles a and b

DNA transfer
replication

cell with
recombined DNA

b

1 Conjugation in bacteria

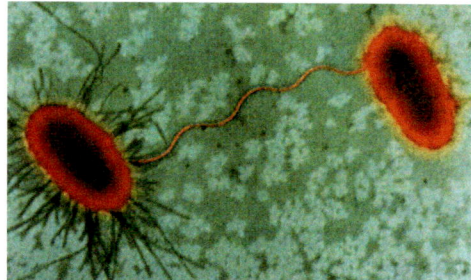

Conjugation in bacteria: cytoplasmic channel

Genetic recombination in prokaryotes

Conjugation
DNA transfer from one
bacterium to another
via a cytoplasmic
channel

Transformation
Uptake and incorpo-
ration of free DNA into
bacteria or other cells

Transduction
Transfer of foreign
DNA into a cell by
using a bacteriophage

Transposition
The movement of DNA
segments within the
genome

Genetic recombination is an important component of sexual reproduction in euka- ryotes. In prokaryotes neither meiosis nor fertilization occurs but, nevertheless, bacte- ria can combine genetic information anew and thus increase the genetic variability of the population. This ensures their ability to adapt to altered environmental conditions.

The direct DNA transfer from one bacterium to another is called *conjugation*. Hereby, direct cell-to-cell contact *(cytoplasmic bridge)* is established between the two bacteria. A well-examined example is the transfer of the *fertility factor (F-plasmid)* of *Escherichia coli* (fig. 1). Bacteria with the F-plasmid are called F$^+$ and without it F$^-$. The F-plasmid contains about 25 genes that are needed to make the cell-to-cell cyto- plasmic channel. It is duplicated prior to transfer: F$^-$ is the DNA recipient ("female"), F$^+$ the DNA donor ("male"). All offspring have the F-plasmid. The F-plasmid (fig. 1a) is temporarily incorporated into the bacte- rial chromosome. These cells are called *Hfr cells* ("*high frequency of recombination*", fig. 1b). When removed from the bacterial chromosome, the F-plasmid can take genes with it that are therefore also transferred during conjugation. The DNA recipient can thus acquire genes for new characters. The cytoplasmic bridge between the bacteria is unstable and DNA transfer can be disrupted by vibrations already. This can be used for genetic mapping: the longer the cell-to- cell contact continues, the more genes are transferred.

The natural ability of bacteria to recombine is used in genetic engineering: During trans- formation, the bacteria take up free DNA from their environment and incorporate it into their genome. This was used by AVERY and colleagues to establish that DNA is the carrier of genetic information (see page 8).

In *transduction*, viruses are responsible for DNA transfer. During the production of *phages* (see info box on page 47) in the host cell, small DNA fragments from the host can be taken up into the phage and transported to other host cells.

Transposition is a process in which DNA segments are moved within the bacterial genome or transferred to a plasmid. These *transposons* are often replicated before transposition. In 1947, B. MCCLINKTOCK discovered that such mobile DNA seg- ments (*transposons*) are also present in eukaryotes. These transposons are called "*jumping genes*".

Tasks

① One hypothesis states that viruses are derived from plasmids. Explain.

② Why is it not appropriate to differentiate between recessive and dominant alleles in prokaryotes (see info box on page 47)?

Bacteria and viruses

The most important findings in molecular genetics come from bacteria, which are prokaryotes. Bacteria are only about 1 — 2 μm long. The cell membrane is different from that in eukaryotes and is surrounded by a strong cell wall. The only cell organelles present in the cytoplasm are ribosomes, which are similar to those of the mitochondria and plastids found in eukaryotic cells. The genetic information in bacteria lies free in the cytoplasm as coiled circular DNA and it is not packed inside a nuclear membrane as in eukaryotes.

The so-called *bacterial chromosome* and the surrounding cytoplasm are called the *nucleoid*. Smaller circular DNA plasmids might also be present, which replicate independently of the bacterial chromosome. Some plasmids can even be incorporated temporarily into the bacterial chromosome. In contrast to viruses, the plasmids have no extracellular stage but viruses might have evolved from plasmids.

Bacteria divide and replicate rapidly in adequate culture media. Prior to each cell division, the DNA replicates. Within a few minutes, a clone composed of daughter cells arises, each with an identical DNA copy. In a 1-ml bacterial culture, we can find several billions of bacteria and among them mutants. In their biochemical abilities, these mutants differ from the wild type cells.

For example, they can resist the effects of antibiotics (*resistant mutants*) or are unable to produce specific substances (*deficiency mutants*). All mutations show in the phenotype because there is no homologous duplication of the genome. The differentiation made in diploid eukaryotes between dominant and recessive alleles is usually not possible in prokaryotes.

Viruses are less than 500 nm in size and are thus even smaller than bacteria. They are mainly composed of a protein coat that contains one or more nucleic acid molecules. Since

viruses are not cellular and lack their own metabolism, they are not considered as living organisms. Viruses that attack bacteria are called *bacteriophages* or short *phages*. There are two different replication cycles in phages.

The *lytic cycle* is typical for many viruses. Attachment to specific receptor molecules on the cell wall of the host bacterium is carried out with proteins of the tail fibres and tail spikes. The hollow tail penetrates the bacterial cell wall by using enzymes and the phage's nucleic acid is injected into the bacterium. The phage genome consists of about 100 genes, which are now transcribed and translated by the "enslaved" bacterium. The phage is *virulent*. One of the first phage gene products is an enzyme that cuts bacterial DNA. The generated nucleotides are used for the replication of the nucleic acid of the phage. The phage genome thus gains complete control over the bacterium. Meanwhile, the synthesis of phage proteins by the bacterial ribosomes is started. Viruses have only limited types of protein and these can assemble independently based on their shape or charge distribution, thereby forming the phage coat, which surrounds the phage nucleic acid. The enzyme *lysozyme* lyses the bacterial cell wall and the bacterium bursts releasing 50 to 200 phages.

In the *lysogenic cycle of temperate viruses*, the phage genome does not become active straight away in the host cell but integrates into the host genome at a specific site. This *prophage* is duplicated with the bacterial genome prior to each cell division and passed on to the daughter cells. The virus is *temperate*. The prophage leaves the bacterial genome and becomes lytic only under special environmental conditions (*induction*).

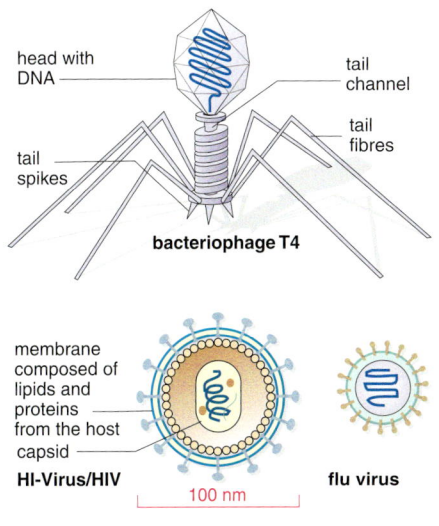

1 Structure of various viruses

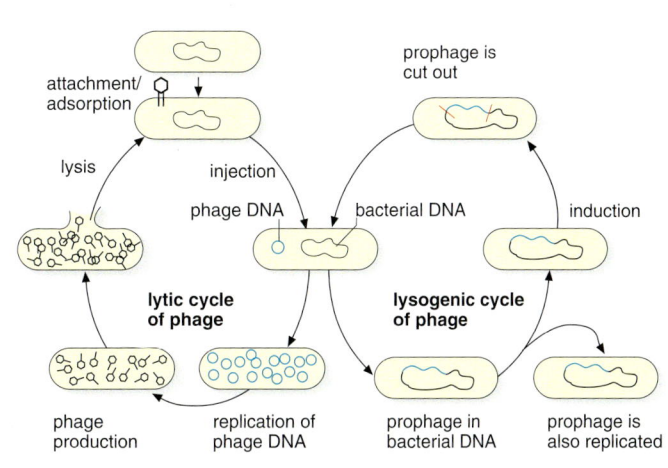

2 Lytic and lysogenic replication cycle of phages

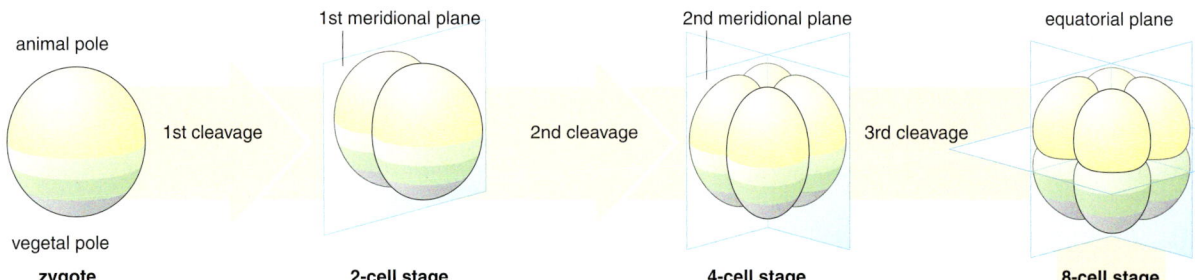

animal pole | 1st cleavage | 1st meridional plane | 2nd cleavage | 2nd meridional plane | 3rd cleavage | equatorial plane

vegetal pole

zygote   **2-cell stage**   **4-cell stage**   **8-cell stage**

From zygote to multicellular organism

Blue whales, sea urchins, fruit flies and humans – they each develop from one single cell called a _zygote_ by countless mitoses. The basis for species-specific development and the composition of an organism is determined by its genes. Environmental factors can have a modifying effect.

The first divisions of a fertilized egg cell have been well examined in sea urchins. The egg cell has little yolk and is polar. The region containing more yolk and pigment is called the _vegetal pole_, whereas the other region side is called the _animal pole_ (see fig.). After fertilization by a sperm, mitoses start in the generated zygote. After the first few cell divisions, the cleavage furrows can clearly be seen running externally from the animal pole to the vegetal pole and also in the equatorial plane. The cleavage furrows divide the cytoplasm of the egg cell into cells of about the same size (_blastomeres_). This type of cleavage is called complete or holoblastic and is typical for cells with little yolk.

The blastomeres divide further without an increase of cytoplasm. Thus, the daughter cells become smaller and receive part of the maternal cytoplasm with its unequally distributed components. All cells have the same genetic set-up but differ in the composition of their cytoplasm. At first, a solid ball forms (_morula_) from which a hollow ball develops (_blastula_). The developing cavity is called _blastocoel_.

After these cleavages, cellular movement and cellular replacement follow, which are together referred to as _gastrulation_. Three cell groups (_germ layers_) form from which the organs later develop. Initially, some cells migrate from the vegetal pole into the cavity of the blastula. These are the precursors of the middle germ layer (_mesoderm_). Subsequently, the vegetal pole invaginates and the shape of the embryo becomes similar that of a ball with a dent. The generated _gastrula_ has a primitive gut (archenteron) with an opening to the outside called the blastopore. The cells of the primitive gut (= digestive tract) become the inner germ layer (_endoderm_). The outer cell layer becomes the _ectoderm_. Skin and nervous system form from the ectoderm. The mesoderm forms connective tissue, muscles and the lining of the coelom. The endoderm forms the epithelial lining of the intestines. The gastrula develops into a planktonic larva (pluteus) that becomes a sea urchin by metamorphosis.

16-cell stage

morula

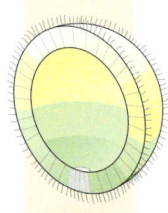

animal pole

vegetal pole
blastula

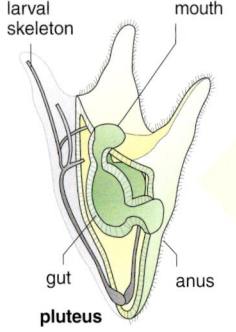

larval skeleton   mouth

gut   anus
pluteus

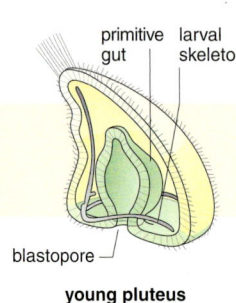

primitive gut   larval skeleton

blastopore
young pluteus

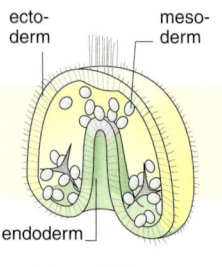

ecto-derm   meso-derm

endoderm
late gastrula

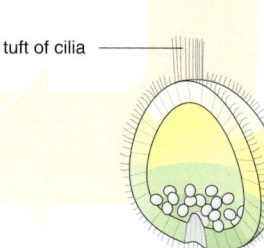

tuft of cilia

early gastrula

Egg and cleavage types

In all multicellular animals, the first steps during development from a zygote to a multicellular organism - cleavage and gastrulation - basically follow the same principles. If we compare the first cleavage stages, we find differences especially in the size and the arrangement of the blastomeres (*blastomere symmetry*).

The different *cleavage types* depend on the amount and distribution of the yolk in the egg cell.

Yolk

Yolk is the nutritive reserve material of an egg cell, e.g. proteins, amino acids, glycogen, lecithin and vitamins stored for the metabolism of the embryo while it cannot take up nutrients.

Yolk is produced in the oocyte during *oogenesis* by helper cells in the female. The helper cells surround the egg cell as a follicle. In other cases, yolk is also secreted by the ovaries.

The yellow of a chicken's egg is the yolk-enriched egg cell, which is surrounded by egg white (albumen) secreted by the ovaries. The egg cell, yolk and egg integument or egg shell together are called the *egg*.

Eggs rich in yolk are often produced by animals that generate only a few eggs, which usually receive intensive *brood care*. Eggs with little yolk are typical for animals that undergo mass reproduction. Mammalian eggs also have little yolk as the mothers carry their embryos inside the womb and nourish them through a *placenta*; the embryos therefore need fewer reserves.

Complete cleavage

Fertilized egg cells with little yolk divide their entire cytoplasm into two cells of about the same size (*blastomeres*). The first two cleavages are meridional, whereas the third is more or less equatorial. It divides the embryo equally into 8 cells of about the same size or unequally into four larger and four smaller cells, as in the sea urchin (see page 48). Further cleavages are *radial*, angularly shifted (*spiral*) or shifted sideways (*bilateral*) to the animal pole. In this way, the later axes of the body are determined.

Partial cleavage

Fertilized egg cells rich in yolk cleave only partially (*discoidal* or *superficial*). In insects, the central nucleus initially divides many times. The daughter nuclei migrate to the periphery of the zygote. Here, the cell surrounds itself with cell membranes (*superficial blastula*). In reptiles and birds, only the *germinal disc* on the animal pole is cleaved (*blastodisc*). The yolk is not divided.

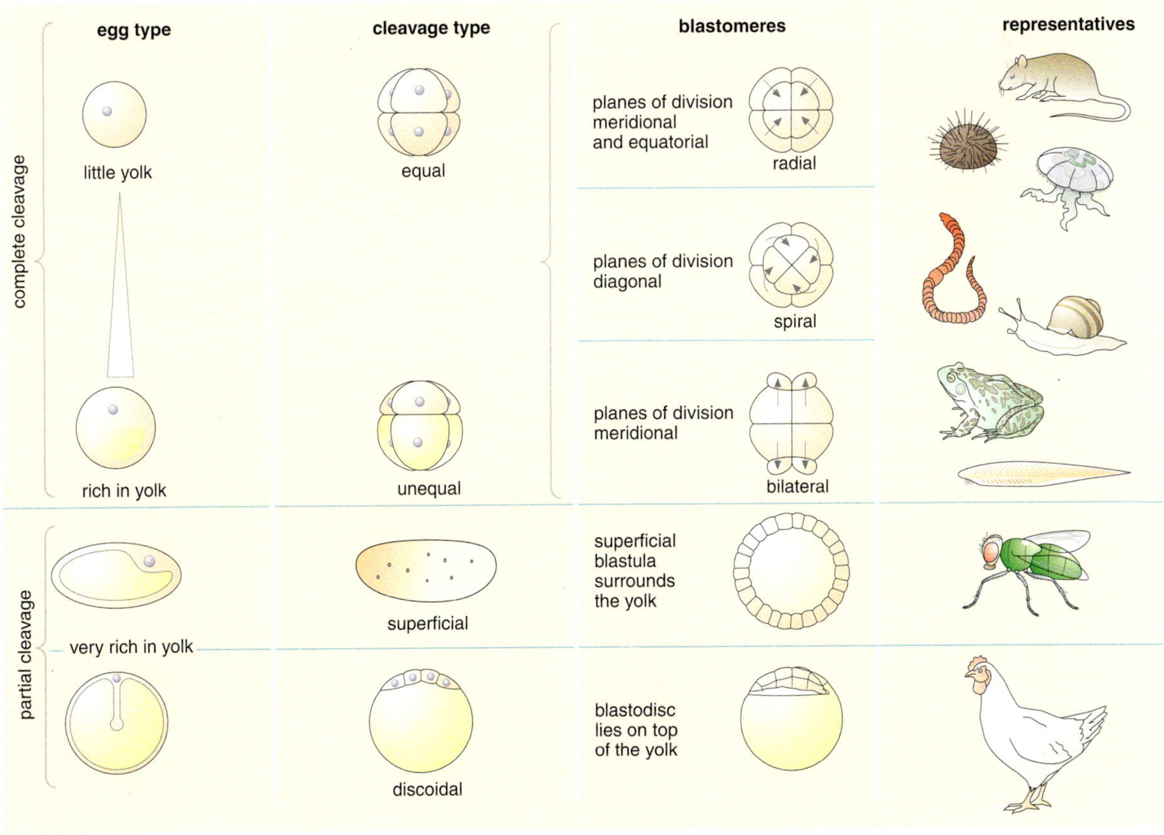

egg type	cleavage type	blastomeres	representatives
complete cleavage little yolk	equal	planes of division meridional and equatorial — radial	
		planes of division diagonal — spiral	
rich in yolk	unequal	planes of division meridional — bilateral	
partial cleavage very rich in yolk	superficial	superficial blastula surrounds the yolk	
	discoidal	blastodisc lies on top of the yolk	

Determination
Restriction of the developmental abilities of a cell

Differentiation
Acquiring the final shape and function

Omnipotent stem cell
("capable of every-thing")
can become a complete organism

Pluripotent stem cells
Such cells can produce all cell types but not a complete organism

Multipotent stem cells
can produce the celltypes of the surrounding tissue

Human stem cell types
Embryonic stem cells: pluripotent
adult stem cells: multipotent

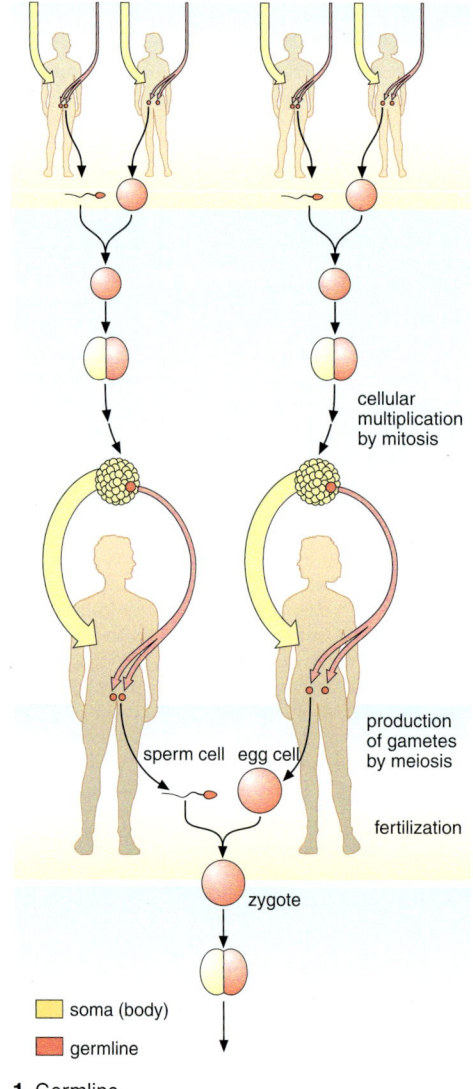

cellular multiplication by mitosis

production of gametes by meiosis

sperm cell   egg cell

fertilization

zygote

soma (body)

germline

1 Germline

In the first few cell divisions, the daughter cells receive the same genetic information but different parts of the maternal cyto-plasm. Concentration differences in mRNA and proteins lead to specific gene expression in the particular cells. Neighbouring cells also influence further development (see page 24). Human cells from the inside of the early gastrula can form all cell types but not a complete organism.

These *embryonic stem cells* are *pluripotent* and probably bear adult stem cells. These cells are able to divide and are responsible for the renewal of tissues in the full-grown organism. *Adult stem cells* are *multipotent* and their cellular environment determines which cell types can be created: adult stem cells of the skin produce only skin cells, stem cells of the red bone marrow produce only blood cells etc. If adult stem cells are removed from their tissue and placed in a test tube, they can be reprogrammed by using various growth hormones. Stem cells from the nervous system can then, for example, develop into muscle cells. Determination means that the genetic information of the cell is not lost but that it is no longer accessed.

Gametes are an exception. They arise by meiotic division from primordial germ cells and become, after fertilization, a new individual in which gametes once again arise in the same way. Cells of the <u>germline</u> are distinguished from cells of the body (*somatic cells,* fig. 1).

Task

(1) The German Embryo Protection Act of 12/13/1990 bans artificial changes in the germline and paragraph 8 states: "Germ line cells, for the purpose of this Act, are all cells that lead from the zygote to the resultant human being and, further, the egg cell from capture or penetration of the sperm cell until the ending of fertili-sation by fusion of the nuclei."
Does this law also apply to stem cells and, if so, to which kind?

Stem cells:
all-rounder and multi-talent

As if following a construction plan, the simi-lar cells of a multicellular embryo develop into specialized cells, for example, muscle, nerve and intestine cells. Development means nothing other than the *differentiation* of cells. Gradually, however, the possibilities of the development of a cell are restricted (*determination*).

The fertilized egg cell of a multicellular organism can become a complete organism. The zygote is an *omnipotent stem cell.* The cells of the first cell divisions are also omnipotent stem cells in humans, and monozygotic twins can result from them.

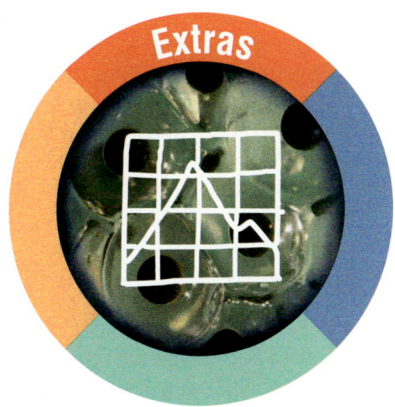

Classic experiments

Amphibians are model organisms for embryogenesis in vertebrates. The front-back axis of the adult animal corresponds to the axis connecting the animal and vegetal poles of the egg cell. The future ventral side is determined by the penetration site of the sperm and, on the opposite side, a differently pigmented area develops. It is called grey crescent. HANS SPEMANN and HILDE MANGOLD (1901/1903) tied various embryos and realized that, in *regulative development*, at least one complete animal is formed, whereas in *mosaic development*, only incomplete structures are formed. Cells are considered to be determined, if they develop, after a transplantation, according to their origin and not to their new location. Furthermore, specific cell groups (*organiser*) have been shown to influence the development of neighbouring cells (*induction*).

Tasks

(1) Read about the reproduction of amphibians and describe the development from zygote until neurula (lefthand side of figure).

(2) Derive from the ligature-experiments the influence that the grey crescent has on development (upper part of figure).

(3) If two blastomeres of a sea urchin embryo are separated at the 2-cell stage, they can become two larvae. From a roundworm embryo treated the same way, no surviving animal develops. Provide the name of the development type and the development potential of the embryos of amphibians, sea urchins and roundworms.

(4) Derive, from the transplantation experiments, the time at which determination occurs in an amphibian embryo (middle and lower part of figure).

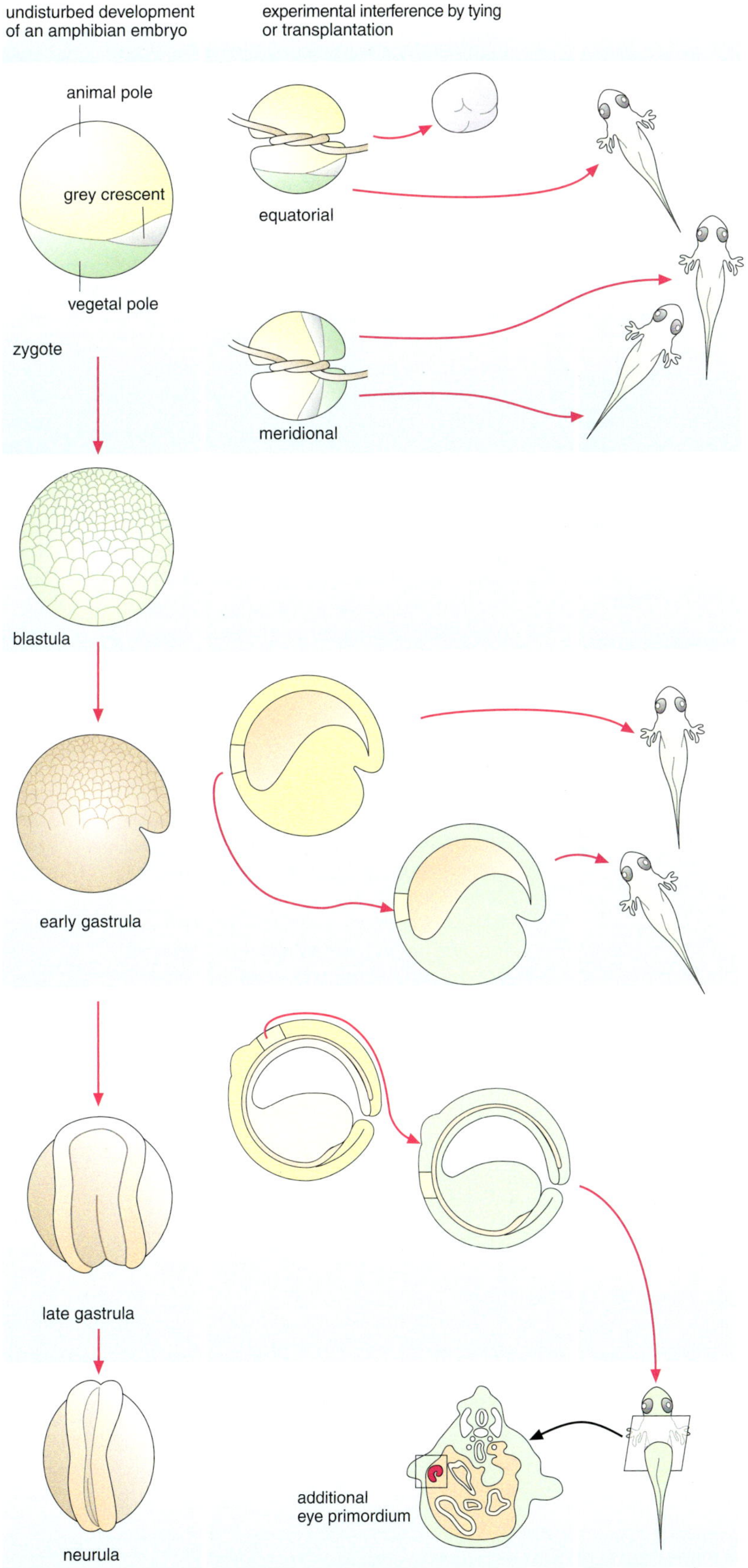

undisturbed development of an amphibian embryo

experimental interference by tying or transplantation

animal pole

grey crescent

vegetal pole

zygote

equatorial

meridional

blastula

early gastrula

late gastrula

neurula

additional eye primordium

Genetics and Immune System **51**

Individual development and gene regulation

In the 18th century, many biologists were still convinced that sperms contained the full-grown organism in miniature form. Today, we know that only the fertilized egg cell holds the genetic information of an individual comparable to a construction plan. However, a construction plan does not make an organism. During embryogenesis, specific genes are activated in a predetermined order. The proteins of these development-controlling genes are DNA-binding molecules, which in turn activate more genes. The formation of an embryonic pattern is based on a hierarchy of activated genes.

Homunculus in sperm

bicoid proteins

↓

activate

↓

gap genes

↓

gap proteins

↓

activate

↓

segmentation genes

Hierarchy of gene activation during the formation of the embryonic pattern

Maternal effects

Essential comprehension about the regulation of developmental processes has been gained from research into *Drosophila melanogaster*. Developmental biologists have tried to understand the mechanisms that lead to segmentation and differentiation during the embryogenesis of a fertilized fly egg.

Drosophila females carry out a type of genetic brood care because they release substances into the oocyte. In the ovaries, maternal cells surround the oocytes (follicle and nurse cells, see fig. 1). The maternal cells located at the front pole of the oocyte "inject" mRNA for a specific protein (*bicoid*) into the oocyte. When, after fertilization, the egg is laid, translation of the maternal bicoid mRNA starts in the front part of the egg. The bicoid protein diffuses through the cytoplasm of the embryo. A concentration gradient is generated. This is possible because, in a fertilized *Drosophila* egg, only the nucleus divides without the formation of cell membranes. The concentration gradient of the bicoid protein can be made visible by using specific antibodies (fig. 2).

Maternal proteins are also produced in this way at the rear pole and on the ventral side of the embryo. Thus, the main body axes of *Drosophila* are defined (between front and back region and also between ventral and dorsal side). The head subsequently forms at the front pole of the embryo, where the concentration of the bicoid protein is highest. Substances such as the bicoid protein, which regulate cellular development and thus, at later stages, the shape of the animal, are called morphogens.

If they are missing, the larva develops incompletely. Egg cells of bicoid-deficiency mutants develop into larvae with no head or thoracic segments. However, the egg cells can be repaired by artificial interference: if the cytoplasm of normal egg cells is injected into the front part, the gradient is re-established and development continues correctly. If the cytoplasm is injected into the middle of the egg cell, a head forms here.

Bicoid proteins cause the production of other proteins, which then in turn activate the so called segmentation genes (see margin).

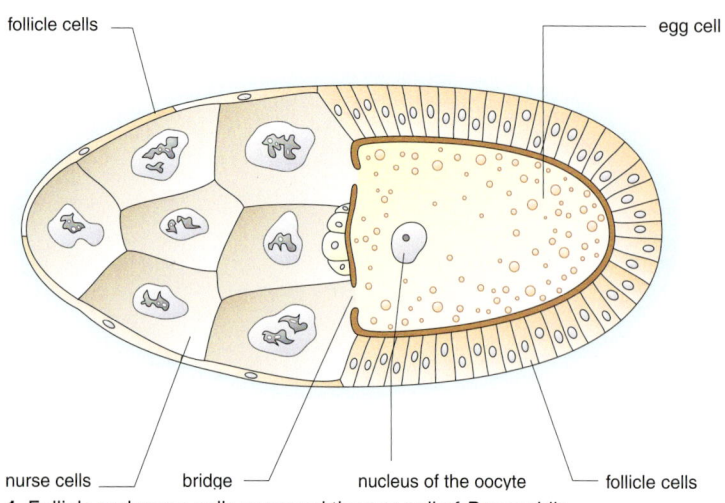

follicle cells — egg cell

nurse cells — bridge — nucleus of the oocyte — follicle cells

1 Follicle and nurse cells surround the egg cell of *Drosophila*

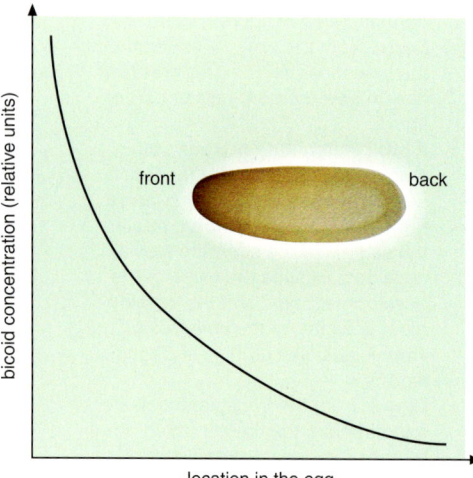

bicoid concentration (relative units)

front — back

location in the egg

2 Concentration gradient of bicoid protein

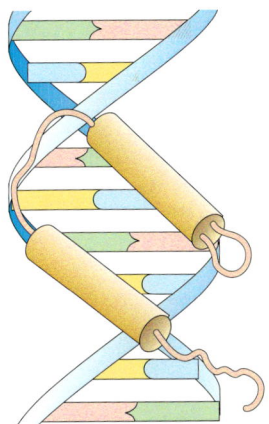

Homeodomains bind DNA specifically

1 The proteins of the segmentation genes

Homeotic genes

These are genes that determine the construction plan of animals by regulating the development of cell groups. The products of homeotic genes (*homeoproteins*) have a section (*homeodomain*) that can bind specifically to DNA.

Figure 1 shows a *Drosophila* embryo at 3 hours after fertilization. Two different segmentation gene products have been made visible using antibodies. The embryo expresses a pattern of 14 stripes in which a brown stripe always follows a grey stripe. The regions between the stripes, without segmentation gene proteins, are not coloured. The future segmentation of the multicellular embryo is thus determined as early as 3 hours after fertilization.

Differentiation of the segments

After a rough arrangement into segments, the so-called *homeotic genes* determine the special composition of each *Drosophila* segment. Mutations that cause, for example, the formation of legs instead of antennae in the head region have revealed the function of these genes.

Homeotic genes resemble each other in one sequence of 180 base pairs. The corresponding amino acid sequence of the encoded proteins forms the so-called *homeodomain* (domain = structural unit) that binds specifically to DNA. The gene products of homeotic genes are regulatory proteins that control the activity of subordinated genes.

Since their discovery in *Drosophila*, homeotic genes have also been discovered in vertebrates and humans. If one transfers a homeotic gene, which is responsible for eye development in mice, to a fly, it produces an additional compound eye. This experimental result indicates that homeotic genes in insects and mammals have the same function.

Task

(1) Mutations in the bicoid gene are phenotypically visible only in the following generation. Explain this statement.

≫info box≪

Fruit fly development

The egg of *Drosophila* has the shape of a long cylinder. During fertilization, the sperm penetrates the oocyte and its nucleus fuses with the nucleus of the oocyte. Within 2.5 hours, about 5,000 nuclei form and float in a common cytoplasm; they later migrate to the periphery. After 22 hours, a segmentated and eating larva has developed from the embryonic stage.

Larva and full-grown fly resemble each other only in their in principle similar repetitive sections *(segments)*. After pupation and metamorphosis, the adult fly hatches. The segments in the head region are fused and contain sense organs such as eyes, antennae and mouthparts. They have differentiated. The three thoracic segments ($B_1 - B_3$, see figure) have appendages such as wings and legs for movement. The abdomen is made up of segments $H_1 - H_8$.

B_1 B_2 B_3 H_1 H_2 H_3 H_4 H_5 H_6 H_7 H_8

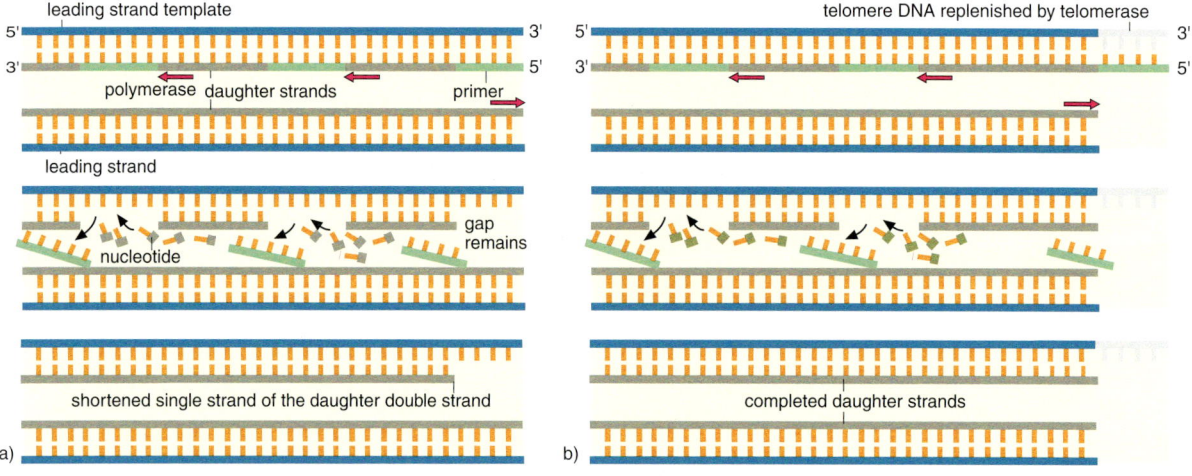

1 Shortening of telomeres during replication: a) without telomerase, b) with telomerase

Labels in figure:
- leading strand template
- 5′ ... 3′
- 3′ ... 5′
- polymerase   daughter strands   primer
- leading strand
- nucleotide
- gap remains
- shortened single strand of the daughter double strand
- telomere DNA replenished by telomerase
- completed daughter strands
- a)
- b)

Aging and death

The skin of the patient is wrinkly, his hair is grey and he suffers from arthritis and eye cataracts. One might think that he is an old man but he is only 12 years old and suffers from the very rare *Werner syndrome*: mutations on chromosome 8 have caused the formation of a changed replication-enzyme complex and development proceeds at fast motion. Does aging follow a genetic programme or is it a deterioration of cells and organs?

Each kind of organism has a specific maximum age that is seldom exceeded (see margin). Nowadays most humans reach old age, but still, not many people live to be more than 120 years; especially long-lived people are often found in one family. Long-lived strains have been bred in *Drosophila* and in the roundworm *Caenorhabditis elegans*. Genes that influence life expectancy and aging have been identified.

Most cells of the tissues in an organism never reach the age of the individual and have to be replaced by new cells (see margin). Cell cultures also perish after 50 to 60 cell divisions. The clock of life ticks with every cell cycle. The counting mechanism is located in the *telomeres* of the chromosomes (see page 15) and are shortened with each replication (fig. 1). When the telomeres are used up, the cell dies of *programmed cell death*. Cells with the unlimited ability to divide (stem cells, tumour cells) possess an enzyme called *telomerase* that re-synthesizes the telo-

mere DNA. In this way, they retain their infinite dividing potential. Whether the age of a cell can also be manipulated by influencing the telomerase without turning it into a tumour cell is unclear.

Nerve and heart muscle cells also age and die, even though they hardly ever divide. For them, shortening of the telomeres is not so important. However, they slowly accumulate pollutants. Reactive oxygen radicals are produced during cellular respiration in mitochondria and cause oxidative damage to nucleic acids, proteins and lipids. If antioxidants obtained from food are not sufficiently present, the damaged cells can die. The cellular-damage clock determines the average survival time of an individual. In metabolically active cells or organisms, pollutants accumulate quickly and deterioration is faster. This could explain the lower life expectancy of small and therefore metabolically more active species in contrast to larger animals.

The aging of cells has genetic and physiological causes and therefore cannot be explained satisfactorily by one theory.

Programmed cell death

The tail of a tadpole is missing in an adult frog. It has not simply dropped off but has "degenerated", which means that the cells have actively degraded in a controlled process and their components have been recycled in the organism. Like cell division, this genetically controlled (programmed) cell death (*apoptosis*) is involved in the formation of tissues and organs. Apoptosis removes surplus, damaged, infected or otherwise modified cells. If the genetic programme for cell death is defective, diseases and deformations can occur.

Today, researchers are certain that, on the one hand, too many apoptotic processes promote stroke, heart failure,

Alzheimer and Parkinson disease, whereas too little apoptosis enables cells to grow without limits and causes tumour or cancer formation. Medical scientists hope to find a way to control cell death and the corresponding diseases pharmacologically.

Tasks

(1) Cell death can be caused by external forces or burns. Explain, using figure 1, how the type of cell death called *necrosis* differs from that of apoptosis.

(2) Necrosis is often put on level with "murder" and apoptosis with "suicide". Explain.

Shaping by apoptosis

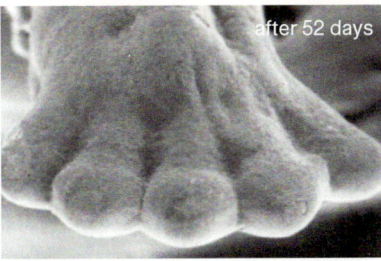

after 52 days

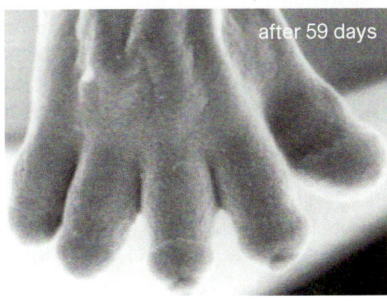

after 59 days

2 Foot of a human embryo

Tasks

(3) Describe the development of a human foot and relate apoptosis to this development (fig. 1).

(4) In many animals, parts of cells and tissues degenerate during development. Give examples in which apoptosis could be important.

The course of apoptosis

Apoptosis is divided into three main phases. In the first phase, an external signal (substances damaging cells, removal of cell contacts, growth factors or hormones) or an internal signal

(mistake during the cell cycle, DNA damage) activates the apoptotic programme. In the affected cell, the regulator protein p53 causes the production of pro-apoptotic gene products so that they outweigh anti-apoptotic substances. P53 "pronounces" the death sentence of the cell by activating degrading enzymes. In the second phase, the cell is destroyed: specific proteases break down proteins and nucleases cleave DNA. In the third phase, membranous vesicles bud off the dying cell, are taken up and digested by macrophages.

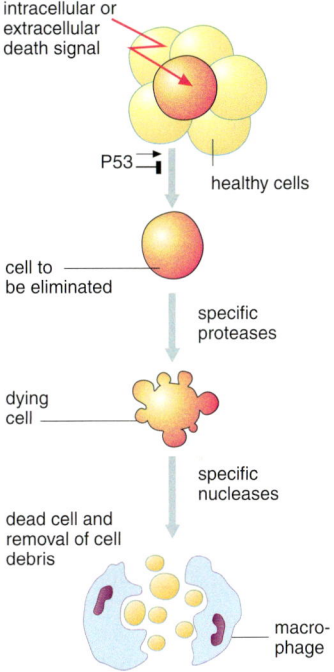

intracellular or extracellular death signal

P53

healthy cells

cell to be eliminated

specific proteases

dying cell

specific nucleases

dead cell and removal of cell debris

macrophage

3 Main phases of apoptosis

Tasks

(5) P53 is encoded by a gene on chromosome 17 in humans. It is called the tumour suppressor gene and inhibits cancer development. Explain this effect.

(6) Viruses causing cancer make a protein that binds p53 thereby inactivating it. How does this improve viral reproduction?

(7) Red blood cells have no nucleus and live only for a short time. Why are old red blood cells eliminated by a special organ, the liver, and not by apoptosis?

Types of cell death

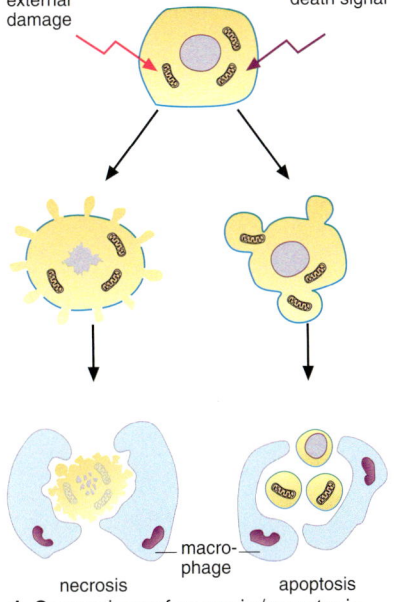

external damage

death signal

macrophage

necrosis

apoptosis

1 Comparison of necrosis / apoptosis

Cancer – the result of misdirected genes?

Tumour
Swelling or lump or growth

Benign tumour
e.g. myoma, polyp

Malignant tumor
Cancer, e.g. carcinoma, sarcoma

Oncogene
A cancer gene

Proto-oncogene
A gene controlling cell division and growth; it can turn into an oncogene by mutation

Other than cardiovascular disease, _cancer_ is the most common cause of death in industrial countries. The term cancer refers to malignant _tumours_ of tissues in which cells divide in an uncontrolled manner and lose their normal function. Tumours displace healthy tissue by over-proliferation and deplete the organism of vital nutrients for their own excessive metabolism. Benign tumours grow very slowly and do not infiltrate the surrounding tissue. They are therefore not necessarily life-threatening. Malign tumours infiltrate the surrounding tissue. Cells can leave the tumour mass and distribute themselves around the body. They can create "daughter tumours" called _metastases_ (see figure 57.1).

Whereas the frequency and time of cell division are regulated in healthy cells, cancer cells are out of control. They divide without limit, do not differentiate and do not belong to a tissue. Molecular biology now offers approaches to explain the development of cancer cells.

As early as 1775, chimney sweepers, who were frequently exposed to tar and carbon particles in soot, were shown to develop cancer more often. Thus, specific substances and also high-energy radiation such as X-rays, UV or radioactive rays can cause cancer. They are _carcinogenic_. Such factors are mutagenic and therefore we can easily assume that mutations in somatic cells (_somatic mutations_) can lead to cancer. In addition, specific viruses can initiate cancer.

The involvement of genes in uncontrolled tumour growth was first realized in cancer-causing viruses. Cancer-causing genes are called _oncogenes_ (Gr. _onkos_ = tumour). Viral oncogenes lead to increased cell division in the infected cells and encourage their own multiplication. The base sequence of viral oncogenes resembles host genes that are responsible for the control of growth and cell division in healthy tissue. If these genes are changed, cellular growth can go out of control. Therefore, they are considered as potential precursors of oncogenes and are called _proto-oncogenes_.

Various mutations can change vital proto-oncogenes into cancer-causing oncogenes (fig. 1): Multiplication (multiple duplication) of the proto-oncogene increases its expression and thereby cell division. Specific point mutations can produce gene products with increased function. Translocation of a proto-oncogene can leave it in the DNA neighbourhood of a highly active promoter (see page 17).

Cancer not only develops if growth-promoting genes are mutated, but also if genes are mutated that usually inhibit cell growth and that as intact molecules hinder tumour formation. These are therefore called _tumour suppressor genes_. Mutations in tumour suppressor genes are connected with various types of cancer, e.g. malignant retinoblastoma in children. Somatic genomic mutations (see page 28) are also thought to be responsible for cancer development.

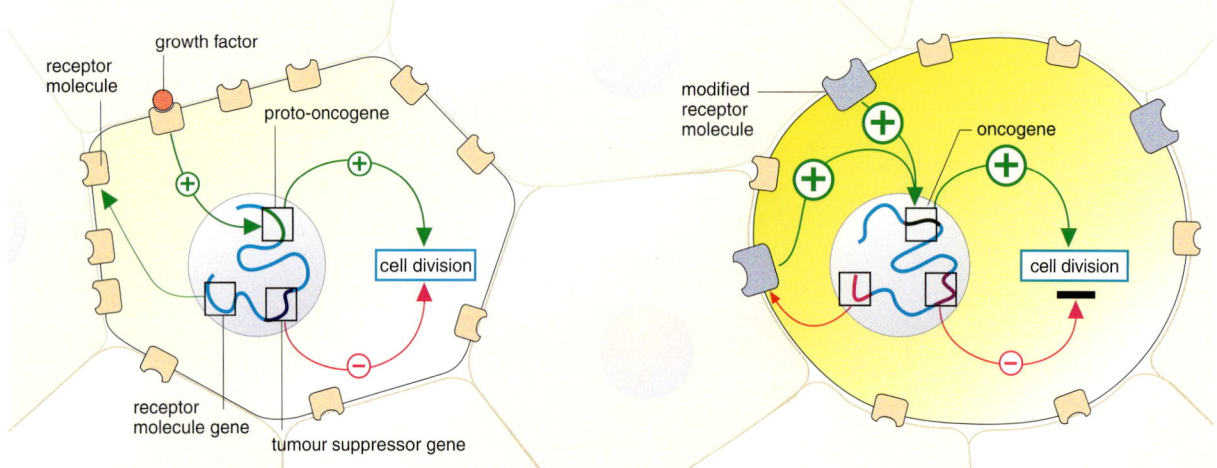

1 Regulation in normal somatic cells (left) and tumour cells (right)

Cancer is usually caused by several mutations. That explains the higher cancer frequency in older patients and an accumulation of cancer in certain families. The *two-hit hypothesis* assumes that at least two mutations are necessary to turn cells into cancer cells: one mutation damaging, for example, a gene regulating cell division and the other one interfering with tissue formation. If one person has already inherited a mutated gene, one further mutation is sufficient to cause cancer in somatic cells. The person concerned is *predisposed*.

The two-hit hypothesis explains the development of colon cancer, the most frequent cancer type in men. In healthy cells, the proto-oncogene on chromosome 12 encodes a membrane-bound receptor protein. Inside the cell, it activates the programme for cell division as soon as growth factors attach to the cell. If a mutation turns the proto-oncogene into an oncogene, the modified receptor protein is always active and continuously stimulates the cell to divide even in the absence of growth factors. The result is a benign growth (*polyps*). The loss of a tumour suppressor gene by a further mutation then leads to cancer (fig. 1).

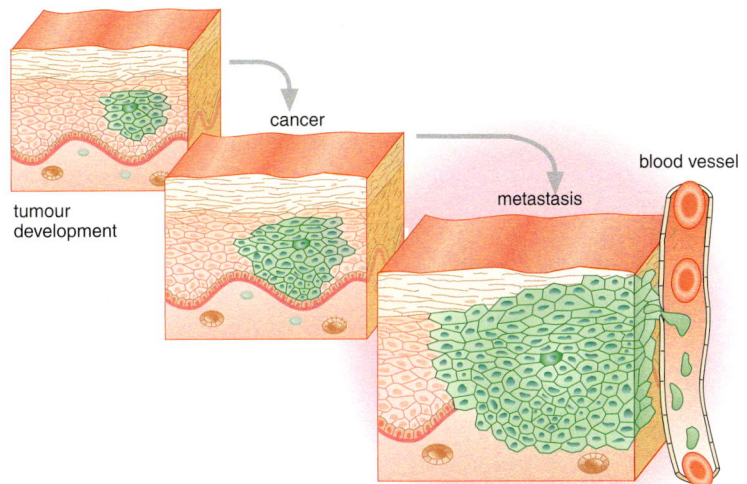

1 Cancer development in the colon

Tasks

(1) The mutation of a proto-oncogene to an oncogene is dominant. Mutations causing a loss of function of a tumour suppressor gene are recessive. Explain this.

(2) Cigarette smoke and alcohol can change the gene on chromosome 17 coding for the regulating protein p53 (see page 55). Explain the connection with the cancer-causing effect of smoking.

(3) Look up further risk factors that can lead to cancer development.

»info box«

Methods of cancer therapy

The diagnosis "cancer" today is no longer a "death sentence". The classic therapy has been refined and thereby enhanced. Better operation techniques help to remove tumours completely before they have time to infiltrate the surrounding tissue or produce metastases. Substances inhibiting cell division (*cytostatics*) can stop the growth of cancer cells. New types of cytostatics fight the tumour cells more specifically and therefore have fewer side effects and better inhibit the development of resistance in tumour cells. Radioactive radiation of tumours is controlled very precisely in the meantime so that fewer normal cells are damaged.

Molecular genetic findings make cancer therapy easier. In particular, genetic diagnostics to identify oncogenes allow early risk analysis in people from families in which specific cancer types occur more frequently. Prevention can be intensified and perhaps a preventive operation can even be carried out. Genetic analyses show the successes of cancer aftercare. In many operated patients, the tumour occurs again if the cancer cells were not removed completely. PCR methods (see page 72) can indicate the presence of single cancer cells and therapy can then be continued until they are gone. Furthermore, the patient can be saved from the unnecessary continuation of therapy.

Other successful therapeutic approaches can be carried out with immunological methods. The immune system recognizes irregular surface markers on cancer cells and destroys them. Cancer can manifest itself only if this system is overloaded. The immune system of the patient can be activated or immune cells directed against tumours can be proliferated in cell cultures and then be put back into the body. If immune cells against cancer cells are linked with strong radioactive substances or strong toxins, tumour cells can be specifically destroyed. Normal cells lacking the tumour surface markers are not attacked. Other medications can block surface markers to which cancer cells attach in the tissue, thereby blocking the formation of metastases.

All methods are more successful, the earlier the cancer is diagnosed. Regular preventive medical check-ups can save lives!

Encyclopaedia

The history of genetics

1866: The Augustinian monk JOHANN GREGOR MENDEL published his "experiments with plant hybrids". From more than 10,000 crossing experiments, he determined ratios from which he derived the laws of inheritance. Their general validity was not recognized by his contemporaries.

1869: The chemist and medical scientist FRIEDRICH MIESCHER isolated "nuclein" from nuclei, which, in 1879, WALTHER FLEMMING calls *chromatin* as viewed in microscopic sections. In 1889, R. ALTMANN proposed the term nucleic acids.

1881: EDOUARD-GÉRARD BALBIANI discovered "nuclear threads" in the nuclei of insect larvae. Only in 1933 could EMIL HEITZ and HANS BAUER establish that these giant chromosomes are indeed chromosomes.

1888: The term "chromosomes" referring to "nuclear loops" observed during cell division was used for the first time by the anatomist WILHELM WALDEYER.

circa 1900: ALBRECHT KOSSEL (Nobel Prize 1910) cleaved nucleic acids chemically and detected phosphoric acid, sugar units and purine and pyrimidine bases as components.

1900: CARL CORRENS, HUGO DE VRIES and ERICH VON TSCHERMARK-SEYSENEGG confirmed independently Mendel's laws of inheritance.

1902: The zoologist THEODOR HEINRICH BOVERI and the cytologist WILLIAM S. SUTTON postulated independently the chromosome theory of inheritance by combining cell biology and genetics.

1905: WILLIAM BATESON, EDITH REBECCA SAUNDERS and REGINALD CURDELL PUNNETT discovered the linked inheritance of genes.

O. T. AVERY

J. G. MENDEL

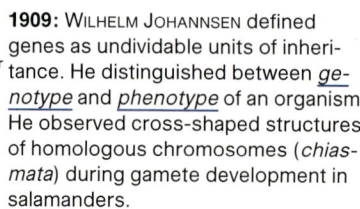

T. H. MORGAN

1909: WILHELM JOHANNSEN defined genes as undividable units of inheritance. He distinguished between *genotype* and *phenotype* of an organism. He observed cross-shaped structures of homologous chromosomes (*chiasmata*) during gamete development in salamanders.

1911: THOMAS HUNT MORGAN (Nobel Prize 1933) realized, from crossing experiments with *Drosophila*, that genes that are inherited together (linked), are located on the same chromosome. Linkage groups can be broken by gene exchange. Based on this, relative genetic maps can be made. The gene distance is given in the unit *Morgan*.

J. D. WATSON

F. H. C. CRICK

1926: HERMANN J. MÜLLER (Nobel Prize 1948) showed that X-rays cause mutations.

1928: FREDERICK GRIFFITH established that genetic information is transfered from one bacterium to another.

1935: MAX DELBRÜCK (Nobel Prize 1969) assumed that genes are fixed molecular units of a substance later recognized as DNA.

1941: GEORGE W. BEADLE and EDWARD TATUM (Nobel Prize 1958) postulated the "one gene-one enzyme hypothesis".

1944: A.M. SRB and N.H. HOROWITZ redefined the term gene: "each gene controls a step within a chain of biochemical reactions."

1944: OSWALD T. AVERY, COLIN MCLEOD and MCLYN MCCARTHY establish that DNA is the carrier of genetic information.

1947: BARBARA MCCLINTOCK (Nobel Prize 1983) described "jumping genes" in maize.

1951: ERWIN CHARGAFF discovered that the 4 bases in DNA are present in a specific ratio to each other (*Chargaff's rule*).

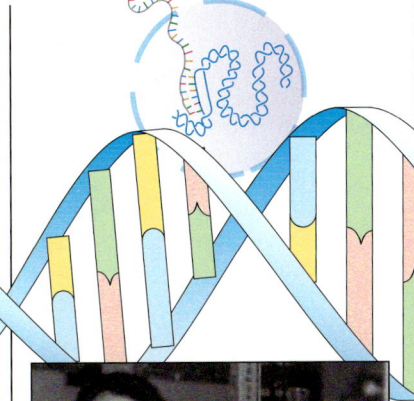

1969: Joseph G. Gall and Mary Loupardue invented a method based on the *in-situ hybridisation* of nucleic acids. This technique allows the localisation of genes on a chromosome.

J. H. Matthaei and M. Nirenberg

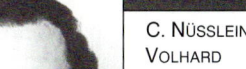

F. Jacob and J. L. Monod

1953: James D. Watson and Francis H. C. Crick (Nobel Prize 1962) postulated the double helical structure of DNA in which the results of Rosalind Franklin, amongst others, are included. She did not receive a Nobel prize because she had died prior to the awarding of this prize (aged 37).

1958: Matthew Meselson and Frank and Mary Stahl demonstrated that DNA replication is semi-conservative.

1959: Claus Pelling showed gene transcription by demonstrating RNA synthesis in puffs of giant chromosomes.

1961: François Jacob and Jacques L. Monod (Nobel Prize 1965) published the *operon model* illustrating the way in which genes are turned on and off. They postulated the "*one gene-one mRNA hypothesis.*"

1965: The work of Marshall Nirenberg (Nobel Prize 1968) enabled Heinrich Matthaei and Severo Ochoa to clarify the genetic code: three nucleotides always define one amino acid.

1970: ·Howard Temin and David Baltimore (Nobel prize 1975) find *reverse transcriptase* in <u>retroviruses</u>. Retroviruses translate their RNA genome to DNA with the help of this enzyme.

1972: Werner Arber (Nobel Prize 1978) discovers restriction enzymes ("enzymatic scissors").

1972: Paul Berg and colleagues (Nobel Prize 1980) built the first recombinant DNA molecule by cutting DNA and sticking the fragments together in a new way.

1973: Stanley N. Cohen, Herbert W. Boyer and colleagues showed that recombinant DNA can direct the production of proteins in living cells.

1975: Georges Köhler and Cesar Milstein (Nobel Prize 1984) succeeded in the production of monoclonal antibodies.

1977: Richard J. Roberts and Philipp A. Sharp (Nobel Prize 1993) discovered the *splicing* reaction in eukaryotes. Introns are removed from the primary transcript (*pre-mRNA*), and the exons are fused to form the mature mRNA still inside the nucleus.

1977: Allan Maxam and Frederick Sanger (Nobel Prize 1980) independently developed effective methods for the sequencing of DNA.

1982: The first genetically engineered drug (*insulin*) became available on the US market.

C. Nüsslein-Volhard

W. Arber

1983: Kary B. Mullis (Nobel Prize 1993) invented the polymerase chain reaction (*PCR*). This technique faithfully copies DNA in a cell-free system.

1985: Alec J. Jefferys, Victoria Wilson and S. L. Thein developed the genetic fingerprint method.

1990: The human genome project started to decode the human genome.

1995: Edward B. Lewis, Christiane Nüsslein-Volhard and Eric F. Wieschaus (Nobel Prize 1995) clarified basic genetic mechanisms of embryonic development.

1996: The genome of a complex organism, the baker's yeast, was completely decoded.

2001: Leland H. Hartwell, R. Timothy Hunt and Paul M. Nurse (Nobel Prize 2001) revealed the key genes controlling the cell cycle in yeast.

2002: Sydney Brenner, H. Robert Horvitz and John E. Sulston (Nobel Prize 2002) used the round worm *Caenorhabditis elegans* to follow the genes that are responsible for programmed cell death (*apoptosis*). Such genes are also active during human embryogenesis and tumour defence.

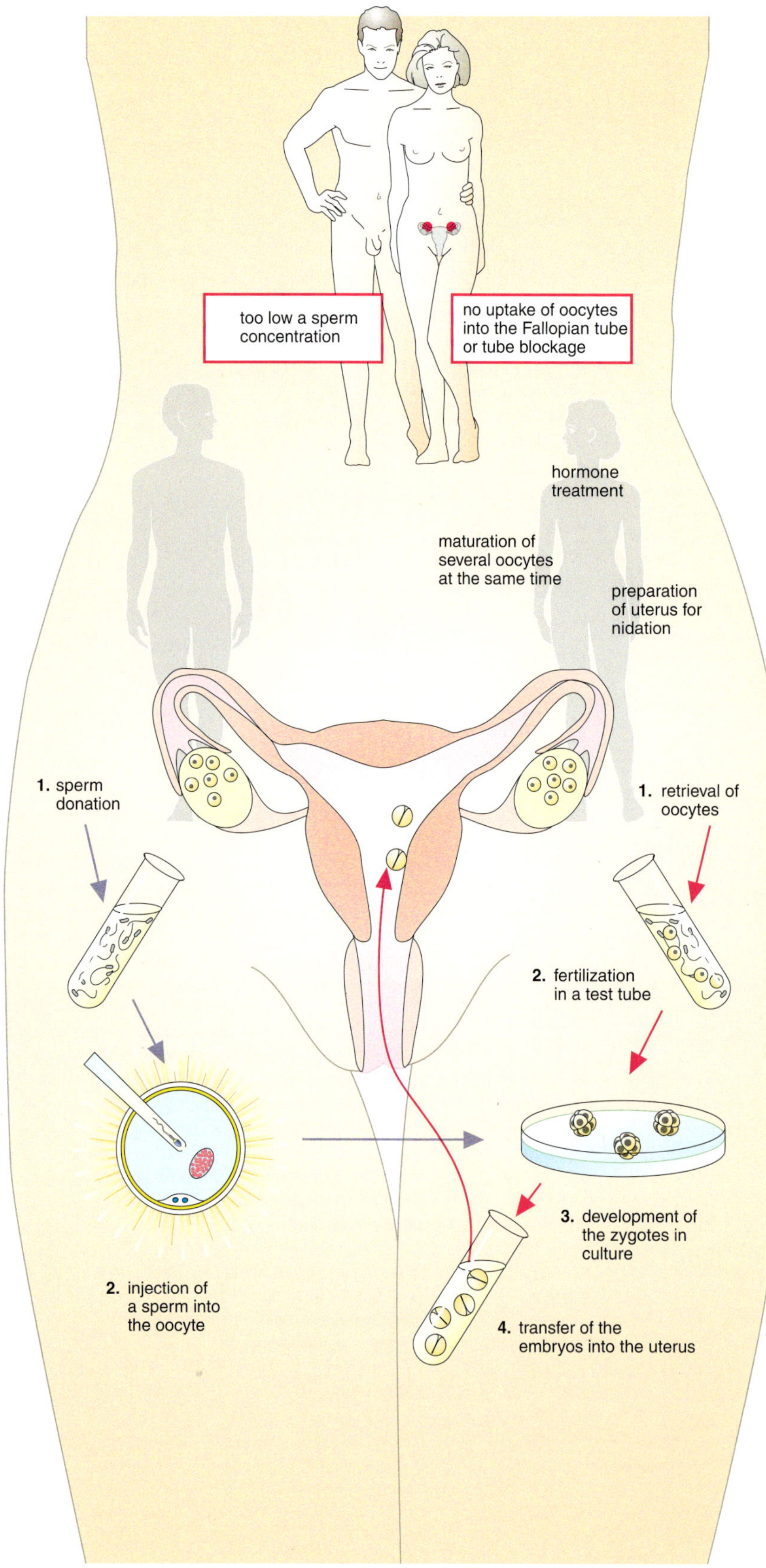

1 Methods of in vitro fertilisation

Labels within figure:

too low a sperm concentration

no uptake of oocytes into the Fallopian tube or tube blockage

hormone treatment

maturation of several oocytes at the same time

preparation of uterus for nidation

1. sperm donation

1. retrieval of oocytes

2. fertilization in a test tube

2. injection of a sperm into the oocyte

3. development of the zygotes in culture

4. transfer of the embryos into the uterus

Reproduction in a test tube

In Germany, about 700,000 couples are unintentionally childless and seek medical help every year. There are different causes for infertility. In women, these are: hormone dysfunction, problems during the uptake of the oocyte from the ovary into the Fallopian tubes or a Fallopian tube blockage. In men, the causes for infertility can include: too low a sperm concentration or non-motile sperms.

In the case of a Fallopian blockage on both sides, fertilization can be performed in a test tube (*in vitro fertilization*, *IVF*). The development of many mature oocytes is stimulated by hormones. The oocytes are aspirated just before ovulation (ultrasound control) from the ovaries and placed into a test tube where they encounter the sperm of the partner. Whether fertilization has occurred is checked microscopically after about 18 hours. The fertilized oocytes are kept in culture until multicellular embryos have developed. These can be transferred into the uterus of the woman through a catheter. The woman has to be pre-treated with hormones so that the uterine lining is ready to host a nesting embryo.

If the concentration of sperm cells in the ejaculate of the man is too low, it is possible to inject single sperms directly into the plasma of an oocyte. This is carried out under the microscope by using a fine glass capillary. The artificially produced zygote develops and is treated as an oocyte fertilized by IVF.

Tasks

1. Use, for example, the internet and find out the legal background to IVF in Germany (possible search terms: "German Embryo Protection Act", third party sperm or ovum donation, surrogacy and surplus embryos).

2. It is possible to determine several genetic features (e. g. sex or dispositions for specific diseases) during in vitro fertilization. This is called *preimplantation genetic diagnosis* (*PGD*). Which laws have been enforced to inhibit the selection of embryos based on specific features (*embryo selection*, see page 71)?

Reproduction technology and cloning

A *clone* is the genetic double of an organism. Clones exist in nature, e.g. cuttings taken from plants by vegetative reproduction, and also in the form of monozygotic twins. Reproductive biology labs have tried to produce clones of farm animals (e.g. sheep, pig, cow) in order to obtain individuals with clearly defined features. They want to prevent the "interfering" recombination of the genetic material during sexual reproduction. The success of cloning, however, has been limited and often hundreds of experiments have to be carried out.

Moreover, cloned animals often suffer from diseases. Clone researchers are trying to determine the origin of the organ damage, immune deficiency and early aging that appear in clones.

The Dolly method

Nuclear transplantation

In 1997, after hundreds of unsuccessful trials, an adult mammal was finally cloned. Oocytes of a sheep were enucleated and the nucleus of the udder cells from another sheep was transplanted into one of the enucleated oocytes. After the product had divided, it was transferred into the uterus of a third sheep that functioned as surrogate mother. The gestation period was successful. However, the *cloned sheep Dolly* showed signs of premature aging (e.g. arthritis).

On the 12/14/2003, Dolly had to be put down because of pneumonia. Sheep usually reach the age of 12 years. Of note, the nucleus used to produce Dolly had been taken from a sheep that was 6 years old, and Dolly was also 6. Whether these diseases are related to cloning is

not yet known. Geneticists, however, had predicted Dolly's premature aging. The telomeres of chromosomes are considered to be cellular clocks of life that become shorter after every division. Dolly's chromosomes were from an older donor sheep and the telomeres were therefore already shortened (see page 54).

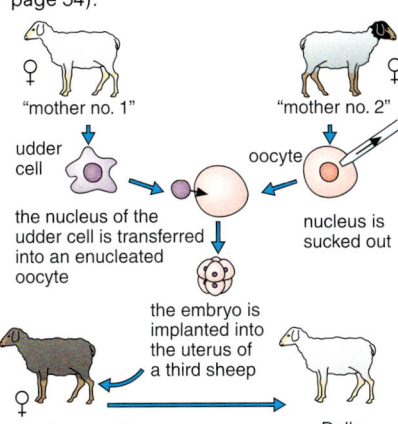

Tasks

1. "Cloning" means the production of a genetic double but also the replication of a gene. Reproduction biology and genetic engineering are often not separated. Why is the cloning of the sheep Dolly not considered a genetic engineering experiment?

2. One of the scientific "fathers" of the cloned sheep Dolly, warned against cloning experiments in humans. There is "abundant evidence that this would be absolutely irresponsible". Explain this.

3. In most countries, it is prohibited to clone humans. Why?

Reproduction stories

The cloned human

(AP news, Washington)
The american researcher JERRY HALL was able to clone human embryos for the first time. In order to avoid ethical problems, he used abnormal embryos that were not able to survive.

A total of 17 embryos of 2 and 8 cells were divided into 48 single cells and surrounded by a protective sheath. The new single cells that were produced this way divided on average three times. According to him, no clone embryo lived longer than 6 days.

Disagreement about frozen embryos

(AP announcement)
In the US, a divorce court has to decide what should happen with frozen embryos. The couple had opted for in vitro fertilization after three childless years.

Of the 10 embryos that had been produced in this way, three were transferred into the woman and 7 were frozen. The woman did not become pregnant. The psychological stress experience at this time was too great for the couple who therefore became divorced.

However, the woman still wanted children. She now wants to have the frozen embryos implanted. The man wants to prevent implantation with the help of the court.

Types of motherhood

According to present law, a mother is the person who has given birth to a child. Since the advent of in vitro fertilization and embryo transfer, three types of motherhood are distinguished: the genetic, the physiological (surrogate mother) and the social mother.

Tasks

4. The manipulation of human embryos is generally prohibited in Germany. Basic research on imported human embryonic stem cells is approved on request. Obtain information about the laws in other countries.

5. Act out a court hearing with different roles in which a divorced couple fights about the the future of their frozen embryos.

6. Reproductive medicine can significantly extend the age at which a pregnancy is still possible. In Austria, for example, the desire for a 61-year-old woman for a child was fulfilled. Here, the oocytes of a young donor were removed and fertilized in vitro with the sperm of the older woman's partner. The embryos were implanted into the uterus of the 61-year old. What social consequences might this have for the parents, the produced children and society?

Methods used in human genetics

Inheritance in humans follows the same laws as in other organisms. *Human genetics* examine the inheritance of blood groups, genetic disorders and disabilities and also possibilities for diagnostics. Classical genetic methods such as family tree analysis are still important instruments and are often the starting point for further examinations. When combined, these different methods can provide a clear picture (fig. 63.1).

Pattern of inheritance analysis

When analysing family trees, the genotype is assumed from the phenotype in order to reveal the pattern of inheritance of the feature. Human geneticists use specific symbols to make up a family tree (see margin).

First, we check whether dominant or recessive inheritance is involved. In dominant inheritance, the presence of one allele (*heterozygosity*) is sufficient to express the feature in the phenotype. A recessively inherited feature shows in the phenotype only in the case of *homozygosity*; although heterozygous parents do not express a feature, their children can express this feature if they receive the corresponding alleles of both parents.

Alleles located on the 22 human autosomes are *autosomally* inherited. In contrast, alleles located on the gonosomes (X and Y chromosome) are *gonosomally* inherited and therefore depend on the sex of the individuals. Gonosomal inheritance shows special characteristics in a family tree: if men inherit a recessive allele on the X chromosome of the mother, then the feature will be expressed, because the second homologous X chromosome is missing (*hemizygosity*). X chromosome recessive inheritance is recognized by a higher frequency of affected men expressing the feature in a family tree. Since women have two X chromosomes, the feature is only expressed in women if both recessive alleles occur together. Heterozygous feature-free women who pass the recessive allele on to their sons are called *carriers*. In X chromosome dominant inheritance, women express the relevant feature twice as often as men. Men expressing the feature always inherit the allele from their mothers and always pass the allele on to their daughters.

Haemophilia A model

The blood clotting disorder, now known as *haemophilia A*, became famous because of its occurrence in European royal families descending from Queen Victoria (fig. 1).

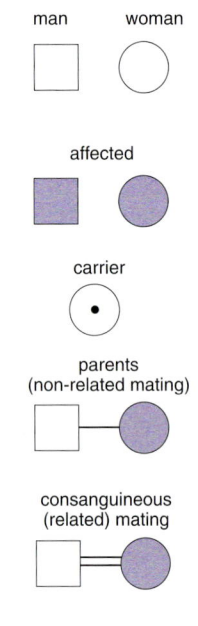

man   woman

affected

carrier

parents
(non-related mating)

consanguineous
(related) mating

siblings

oldest   youngest

dizygotic twins

monozygotic twins

Family tree symbols

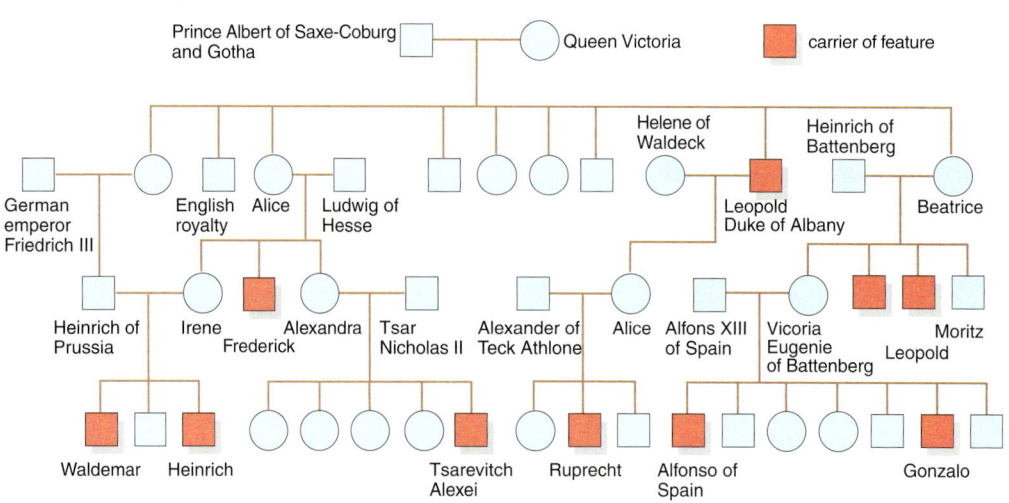

Prince Albert of Saxe-Coburg and Gotha   Queen Victoria   carrier of feature

Helene of Waldeck   Heinrich of Battenberg

German emperor Friedrich III   English royalty   Alice   Ludwig of Hesse   Leopold Duke of Albany   Beatrice

Heinrich of Prussia   Irene   Frederick   Alexandra   Tsar Nicholas II   Alexander of Teck Athlone   Alice   Alfons XIII of Spain   Vicoria Eugenie of Battenberg   Leopold   Moritz

Waldemar   Heinrich   Tsarevitch Alexei   Ruprecht   Alfonso of Spain   Gonzalo

1 Family tree of the blood clotting disease (haemophilia A) in the European royal line

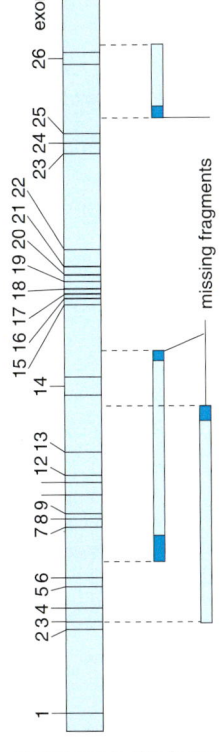

factor VIII gene (haemophilia A)

photo-sensitive pigment (red-green colour blindness)

Gene loci on the X chromosome

exons

missing fragments

Deletions in the factor VIII gene seen in haemophilia A patients

Healthy women pass the disposition for the disease on to their sons whose fathers are healthy. Haemophilia A occurs with a frequency of 1:10,000, which means that, of every 10,000 male newborns, one is affected. This does not depend on ethnic origin.

Biochemical explanation: Haemophilia A patients lack a protein of the blood clotting cascade (*factor VIII*). This glycoprotein contains 1997 amino acids. In the body fluids of healthy individuals, the concentration of factor VIII is about 0.5 to 1.0 mg/100 ml. About half of all haemophilia A patients have less than 1 % of this concentration.

Some patients have 30 % of the normal concentration but, because of differences in the amino acid sequence, the gene product is not functional and blood clotting does not occur. Heterozygous carriers of the haemophilia A gene have a lower factor VIII concentration than healthy people but the concentration is sufficient to ensure blood clotting. In the past, haemophilia A patients were given the plasma of blood donors. Unfortunately, this added the problem that pathogens (e.g. hepatitis viruses) were often also transmitted to the haemophiliac. Since factor VIII can be produced by gene engineering, sterile medication is now available in adequate amounts.

Cytogenetic explanation: The recessive allele for haemophilia A, which is located on the X chromosome (see margin), is rare in the human population. Women function as carriers and pass the allele on to their child-

ren. These women do not suffer from the disease because they have two X chromosomes, one of which contains the dominant allele (the same is true for their daughters). The male genotype contains only one X chromosome received from the mother. The sons inherit the Y chromosome from their fathers but this chromosome does not contain any genes relevant for the haemophilia system. If a man inherits the haemophilia allele from his mother, he will therefore become ill because he lacks the dominant allele.

Molecular biology explanation: The factor VIII gene has 2,351 codons distributed over 26 exons (see margin). In 50% of cases, the severe familial forms are based on deletions or inversions of large gene segments.

Tasks

① Analyse the family tree of albinism, inherited brachydactyly (shortness of fingers) and red-green colour blindness. Which pattern of inheritance occurs respectively? Determine the genotype of as many individuals as possible.

② Why does haemophilia A occur more rarely in women than in men?

③ Which individuals act as carriers in haemophilia A (fig. 62.1)? Explain.

④ Genetic diseases are examined with various scientific methods. Make a table with regard to haemophilia A and list the findings of classic genetics, cytogenetics, molecular biology and biochemistry.

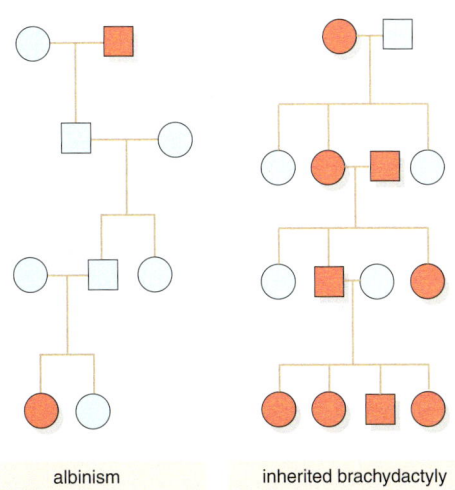

albinism

inherited brachydactyly

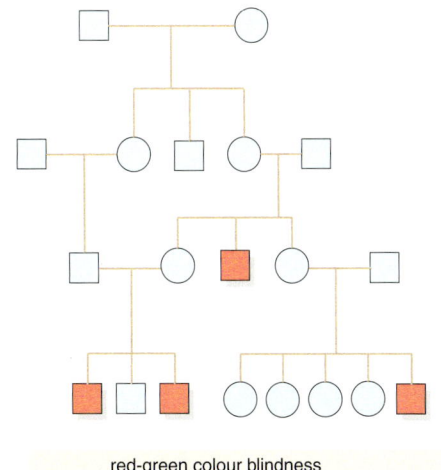

red-green colour blindness

1 Family trees of various hereditary diseases

Patterns of inheritance

Often the relationship between genotype and phenotype is only revealed when a mutation of the gene changes the gene product and its function. Some genetic diseases that are based on the modification of a single gene are said be to of *monogenic* origin. For basic genetic research and medicine, these monogenic diseases are extremely helpful.

Autosomal dominant inheritance

So-called *Marfan syndrome* manifests itself as defects in the skeletal system (arachnodactyly ["spider fingers"] and chest deformations), the eye (lens damage) and the heart. The disease frequency is 1 : 10,000, which means that, of every 10,000 newborns, one is ill, independent of sex. The responsible gene (*FBN1*) is located on chromosome 15. It codes for a calcium-binding protein that is part of the elastic fibres in connective tissue. This explains the widely distributed damage caused by mutations in the FBN1 gene. The FBN1 gene product is not a particularly large protein. Most organ damage is caused by mutations in the calcium-binding domains of the protein. The calcium-binding amino acid cysteine is missing in a key position and the connective tissue cannot form as usual. The presence of one defective allele causes the production of defective proteins. These are incorporated together with intact proteins into the connective tissue, which is now also defective. Therefore, the allele is dominant.

Autosomal recessive inheritance

Phenylketonuria (PKU) is the most common amino acid metabolism disorder in humans (regional frequency up to 1 : 2,500). The amino acid *phenylalanine* (Phe) is taken up with food and cannot be converted to the amino acid *tyrosine* (Tyr) because of an enzymatic defect (see figure). Phe accumulates up to 30 times of its normal concentration and is converted to toxic metabolites (phenylketones). These affect, for example, the formation of the myelin sheaths around nerve cells.

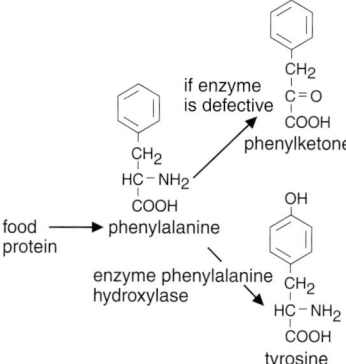

Severe brain damage, paralysis and cramps are the results. Newborns are routinely screened for increased Phe blood concentration by using mass spectrometry tests. If the result is positive, a diet low in Phe and Tyr is given and the brain develops normally. PKU has an autosomally recessive pattern of inheritance, which means that, if a mutated allele and a normal allele are present, enough enzyme is produced to degrade phenylalanine. The gene coding for the enzyme phenylalanine hydroxylase is located on the short arm of chromosome 12. Seventy different mutations of this gene are known to cause PKU.

Gonosomal inheritance

X chromosomal dominant inheritance is rare and a special case is rickets. In vitamin D resistant *rickets*, the bones are soft and do not harden because of a calcium deficiency. This results in skeletal deformations. Because of an enzymatic defect, the hormone calcitriol is not synthesised from the precursor vitamin D3. Therefore, vitamin D3 medication does not help. However, calcitriol can be produced artificially and given to the patient to re-establish the normal phenotype.

Calcitriol allows the uptake of calcium in the small intestine and its deposition in the bones. The coding gene is located on the short arm of the X chromosome.

Complex inheritance

In monogenic inheritance, three phenotypes can be found in the population: people carrying a specific feature, people not carrying it and people being only slightly affected. However, some patterns of inheritance do not exhibit these three groupings. If a feature occurs with small variations, it is said to show *continuous variation*. This is often based on the interaction of many genes (*polygenetic*) whose alleles do not have such a strong effect as that in monogenic inheritance. In polygenic inheritance, every single gene is ruled by Mendelian laws and the inheritance of each of these genes can be dominant, recessive, autosomal or gonosomal. The variation in most continuous features is not only based on the genetic background but also on gene-environment interactions and is therefore said to be based on *multifactorial inheritance*. Examples for this are skin colour, body length, body mass or diabetes mellitus.

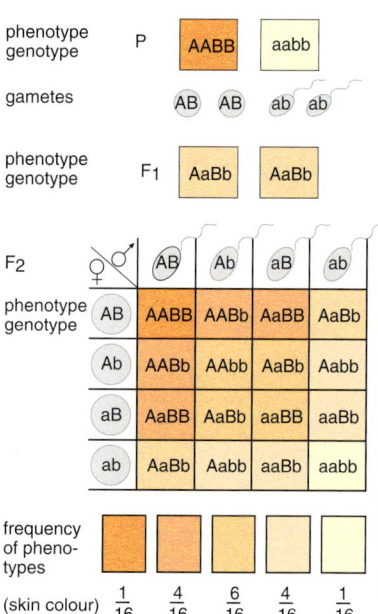

Classic human genetics

Statistics and the blood group system

In humans, more than 100 different blood group systems are known. The ABO system was discovered in 1901 by KARL LANDSTEINER and comprises 4 different phenotypes: A, B, AB, O. The inheritance of blood groups was first explained by dihybrid inheritance in which the blood group O has the genotype aabb. In 1925, the mathematician FELIX BERNSTEIN proposed a monohybrid inheritance with three different alleles (A, B, O) of which every person only contains two alleles in their genotype.

blood group	frequency of genotype in Central Europe		possible gametes
A	AA	31%	only Ⓐ
	A0	11%	Ⓐ or Ⓞ
B	BB	1%	only Ⓑ
	B0	14%	Ⓑ or Ⓞ
AB	AB	6%	Ⓐ or Ⓑ
0	00	37%	only Ⓞ

The alleles A and B are both dominant over the allele O. If A and B are present together, they both affect the phenotype (AB, *codominant alleles*). If more than two alleles exist for a gene, they are referred to as multiple alleles. The hypothesis of multiple alleles was confirmed by BERNSTEIN's analyses of family trees.

He analysed about 3,000 children who had at least one parent with the blood group AB. He found only 13 children with blood group O.

Tasks

1. Design a Punnett square for the features of the ABO blood group system.
2. A dihybrid inheritance pattern with one parent having the blood group AB and hypothetical genotype AaBb should have led to a larger number of children with the blood group O (genotype: aabb) than BERNSTEIN observed. Explain this (3rd Mendelian law).
3. BERNSTEIN could trace all 13 children with the blood group O back to unclear fatherhood. Without this knowledge, his model could have not been firmly established for the ABO blood group system. Explain why a child born with the blood group O cannot have a parent with the blood group AB.

Twin research

The genetic part of the expression of a feature can be determined by comparing twins, even if the respective genes are not known. Twin research is especially important for the analysis of complex multifactorial features. *Monozygotic twins* develop if early embryos, up until the 4-cell stage, divide into two daughter individuals. Monozygotic twins are genetically absolutely identical.

Dizygotic twins are genetically no more closely related than normal siblings.

If monozygotic twins resemble each other more strongly in one feature than dizygotic twins, then the feature is considered to be of genetic origin.

However, if the difference in the expression of the feature is the same for both twin groups, then the expression of the feature depends on the environment. When pairs of monozygotic twins who have grown up separated from each other are compared, features can be identified that are weakly or strongly influenced by environmental factors.

Task

4. Interpret the information given in the tables. Which diseases or features might have a genetic cause?

Disease	Match in %	
	Monozygotic twins	Dizygotic twins (same sex)
Whooping cough	96	94
Appendicitis	29	16
Tuberculosis	69	25
Diabetes	84	37
Bronchial asthma	63	38
Same tumour type	59	24
Stroke	36	19

Examined twin groupe	Mean difference in		
	Body length (cm)	Body weight (kg)	IQ points
Dizygotic twins	4.4	4.4	8.5
Monozygotic twins (reared separately)	1.8	4.5	6.0
Monozygotic twins (reared together)	1.7	1.9	3.1

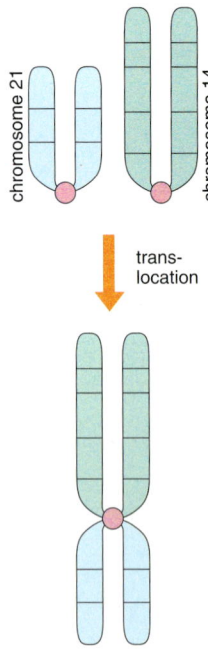

meiosis I

meiosis II

a oocyte forms
that contains two
chromosomes 21

**formation of "free"
trisomy 21**

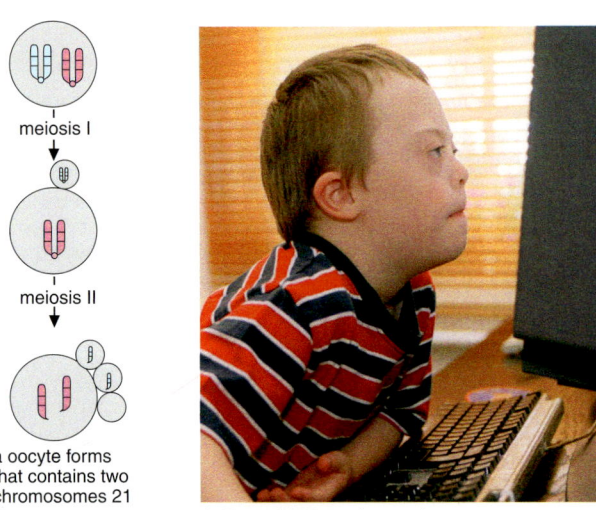

1 Boy with Down's syndrome

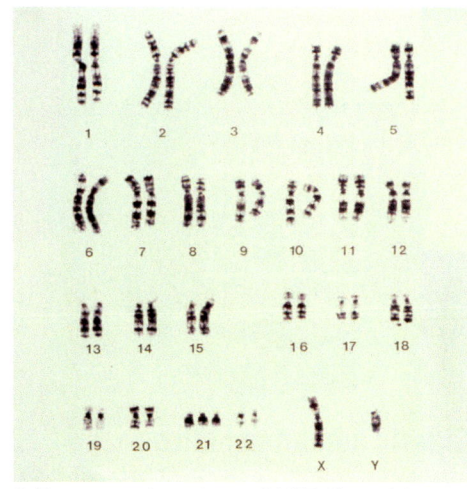

2 Karyogram, free trisomy 21 (47, XY, +21)

Down's syndrome

Syndrome
A complex disease
that is definded by a
specific combination
of symptoms

In 1886, the English paediatrician JOHN LANGDON-DOWN described, for the first time, the symptoms of the *Down's syndrome*: body length below average, shorter fingers and toes, round head, eyes with an angular skin fold on the inner corner of the eye. In addition to these visible features, i. a. heart defects, susceptibility to infections, mental retardation and a shorter life expectancy occur. Down patients have leukaemia and Alzheimer disease more often than is found in the general population.

In 1959, the cytogenetic cause of Down's syndrome was discovered. The karyogram of Down patients contains 47 chromosomes with chromosome 21 being present three times (*trisomy 21*). The reason is often a mistake in the distribution of the chromosomes during gamete development: If the homologous pair of chromosome 21 is not separated in meiosis II (*nondisjunction*) in one parent, two chromosomes 21 are present in the gamete and, during fusion with a normal gamete, a zygote develops that has three copies of chromosome 21 (47, XX or XY, +21). Even though chromosome 21 is the smallest human chromosome, the trisomy strongly affects the phenotype. The gene balance is disturbed.

On average, one child in 700 newborns has Down's syndrome. However, the risk increases with the mother's age (1 : 1,500 in a 20-year-old and 1 : 30 in a 45-year-old mother). In humans, the oocytes remain in meiosis I from birth, with meiosis II (see

page 37) only starting when ovulation occurs. This lengthy storage time is probably the reason that oocytes from older women are more susceptible to defects. In 97 % of all trisomy cases, a false distribution occurs during meiosis and the karyogram of the parents is without pathological findings. "Free" trisomy 21 (fig. 2) can be determined in cell material from the embryo in the amniotic fluid by using prenatal diagnostics (see page 71).

Of the karyograms of Down patients, 3–4 % show an additional chromosome 21 that is fused with chromosome 14 (see margin). This so-called *translocation trisomy 21* can occur if one parent carries the fusion product. During meiosis, chromosome 21 can "piggyback" on chromosome 14 and pass into the gamete. Instead of chromosomes 14 and 21, the gamete contains 14 + 21 and 21. After fertilization, the zygote contains an additional chromosome 21. In contrast to the so-called free trisomy 21 (see above), the translocation can be seen in the karyogram of the parents and the risk does not depend on their age.

Tasks

(1) Gather information about the production of karyograms (see page 15).
(2) Free trisomy 21 can also be based on the abscence of separation of the chromosomes (nondisjunction) in meiosis I. Create a diagram similar to the one in the margin figure.

chromosome 21

chromosome 14

trans-
location

fusion
chromosome

**formation of translo-
cation trisomy 21**

The human genome

The base sequence of the human genome was decoded to a large extent within the *Human Genome Project*. The information about the base sequences of many individuals was collected in databases, made anonymous and released for public access. A comparison of the sequences has shown that the largest part (99.9 %) of about 3 billion base positions is identical in the human genome of people with the same sex. Thus, the remaining part is responsible for the genetic differences between individuals. Nevertheless, this part still contains 3 million sequence differences (= 0.1 % of 3 billion). The greatest part of these differences is composed of the exchange of single bases (*point mutations*) called SNPs (*single nucleotide polymorphisms; pronounced "snips"*). SNPs can occur within genes and also in non-coding DNA segments. Humans can be homozygous for a specific SNP (e. g. T/T, C/C) or heterozygous (e. g. T/C).

Today, methods for automatic sequencing make it possible to examine thousands of samples for SNPs. SNPs have been correlated with diseases of unclear genetic background. This is especially true for multifactorial diseases such as asthma or Alzheimer's disease.

In order to determine such correlations, the SNP pattern of patients is compared with that of healthy individuals. In a second step, the mechanisms that lead to the disease must be analysed. *Functional genome analysis* is engaged with this demanding task. If the SNPs are located in coding DNA segments, the gene products are analysed. However, if they are located in non-coding segments, two possibilities exist: either gene regulation is affected or certain SNPs are coincidently located close to alleles that are related to the cause of a disease and, being physically linked to them, are therefore inherited together with them. For example, several SNPs known to lie on the long arm of chromosome 10 occur close to alleles that are commonly found in Alzheimer's patients. These SNPs can serve as genetic markers to reveal candidate genes for Alzheimer's disease.

No individuals other than monozygotic twins have absolutely identical SNP patterns. Therefore, they also play an important role in crime investigations by providing "genetic fingerprints", especially as, today, even minute traces of DNA are sufficient for accurate analysis.

Every dot on this SNP map of the human chromosome 20 represents 25 different SNPs.

»info box«

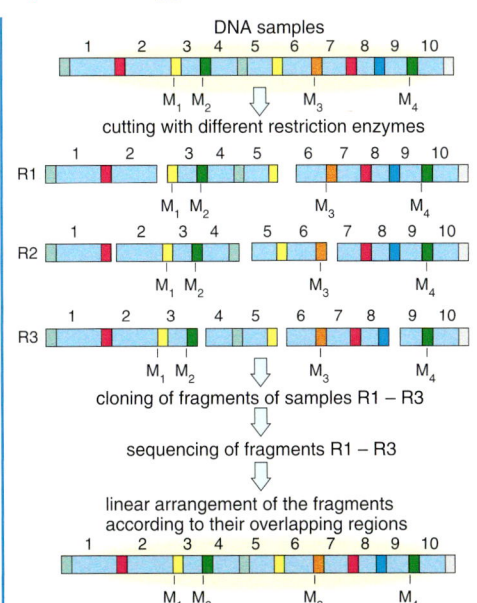

DNA samples

cutting with different restriction enzymes

cloning of fragments of samples R1 – R3

sequencing of fragments R1 – R3

linear arrangement of the fragments according to their overlapping regions

Sequencing the human genome

Human DNA samples are cut by using various restriction enzymes (R1 – R3). Such genetic "scissors" cut the DNA at specific sites. The fragments of the different samples are copied (cloned) and sequenced (see page 73). The fragments are arranged according to the overlapping regions of the base sequence. The continuous base sequence is thus determined.

In order to control whether the fragments are arranged correctly, marker pairs are used (M1, M2, M3, M4). These are DNA segments with a known sequence. Marker pairs that are located on one fragment must also be neighbours on the chromosome. Their proximity can be determined from the frequency with which the pairs occur together on different fragments (see page 43).

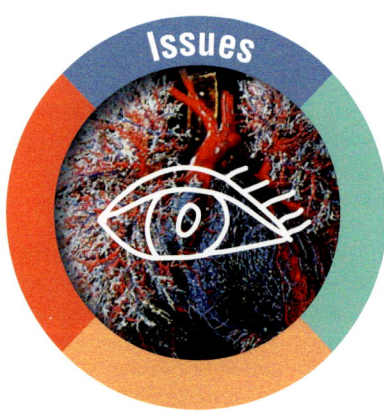

Cystic fibrosis (mucoviscidosis)

Every year, several hundred children are born with *cystic fibrosis* (*CF*, also mucoviscidosis), a hereditary disease. Thick mucus blocks vital organs and a persistent cough, breathlessness and severe digestive problems are the result of this exocrine gland disorder. By using medication that loosens the mucus, inhalation therapy and a special diet, the quality of life of these individuals has improved but there is still no cure. Life expectancy is lowered. In 1948, doctors found that the sweat of CF patients contains too much salt (NaCl). Even today, the NaCl sweat test is the most common method for diagnosing CF. In the 1980s,

CF patients were shown to have faulty chloride ion permeability in the cell membrane of mucous membrane cells. This is caused by a chloride channel that does not open. Now that this correlation is understood, there is hope that drugs can be developed that cause the opening of the channel and thereby remove the symptoms of CF.

State the conditions in which a recessive hereditary disease can persist in a population.

Inheritance pattern of CF

CF follows an autosomal recessive type of inheritance.

It is possible to determine, by prenatal diagnostics, whether a foetus is homozygous for the disease. The technique is safe to 99 %. What problems arise for the parents in the family tree shown below?

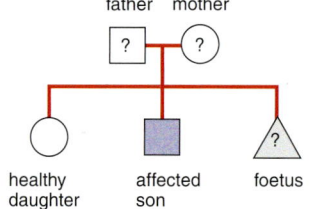

father   mother

healthy daughter   affected son   foetus

Classical therapy

A classical method for providing relief for a CF child is careful pounding on the chest and back. This therapy lasts about 90 min per day. It should help to loosen the thick mucus in the respiratory tract and to remove it. Inhalation therapy is also helpful.

Gather information about other classical therapies of CF (treatment of symptoms).

Coping strategies

"I remember George well. He was a calm pale pupil. He sat in one of the back rows in my specialist biology course. His schoolwork wasn't bad, but I had the feeling that he could have achieved more. His classmates and teachers thought that his persistent coughing was because he had a chronic cold. Now and then, he was absent for a long time during the winter. Being a passionate hobby angler, he investigated dragon fly larvae in his biology project. Three and a half years after his Abitur, I saw his death notice in the newspaper. Accident? Drugs? These thoughts passed through my mind. The school counsellor knew the answer. Years ago, George had told only him that he had cystic fibrosis. He didn't want his classmates or teachers to know."
(report of a biology teacher)

"During my last attack, I got a taste of what it feels like to suffocate" describes 25-year-old Dagmar Ziegler, a student of educational theory. "I know that my disease will get worse but I usually try to forget about it."
(Stern, Hamburg 1989)

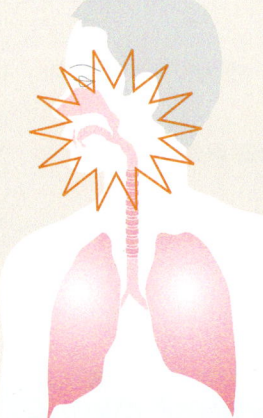

"This is me. There are times when I don't know whether I should laugh or cry. On the one hand, I know that CF is an incurable disease, that my life expectancy is significantly lowered and that CF requires intensive therapy. On the other hand, I receive drugs that help me to get back into shape and that give me a certain amount of security. I have decided on the following basic principles for my life: less work, more therapy and more pleasure. In addition to my part-time job as a type setter, I make sure I have enough time for my inhalation and breathing therapies and for my hobbies of table tennis and the writing of sports articles. This makes it is possible to live well with cystic fibrosis. "
(Klaus, 26 years old)

How do the CF patients in these three examples cope with their disease? Please give your opinions.

The CFTR gene

By 1990, CF was known to show autosomal recessive inheritance and the mutated gene was even known to be located on the long arm of chromosome 7. However, the function of the gene was still unknown. It is a relatively large gene of 250,000 bases in length and contains 27 exons. Only the analysis of the gene has led to conclusions about its function.

It codes for a protein of 1,480 amino acids that, after synthesis, is transported to the cell membrane and is incorporated into it. The gene is called CFTR (*cystic fibrosis transmembrane conductance regulator*). The protein functions as a channel for chloride ions and is opened by ATP. The most common mutation (about 70 % of all cases) is a deletion of three bases in exon 11. Thus, one amino acid (no. 508) is missing. The faulty protein cannot carry out its function.

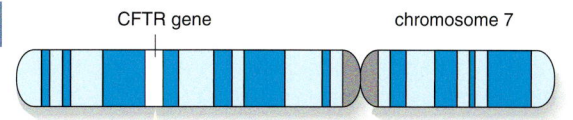

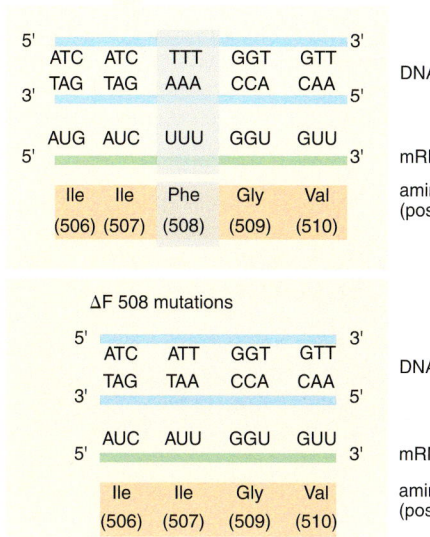

The CFTR gene and the ΔF508 mutation

Meanwhile, more than 400 mutations are now known that lead to different forms of this disease with differing severity. The most severe forms are caused by *nonsense, frame shift or splice mutations* that cause the development of a completely altered and thus non-functional protein. Less severe mutations are *point mutations* that inhibit the correct opening of the channel.

To what extent are monogenic diseases like CF easier to treat with gene therapy than polygenic diseases?

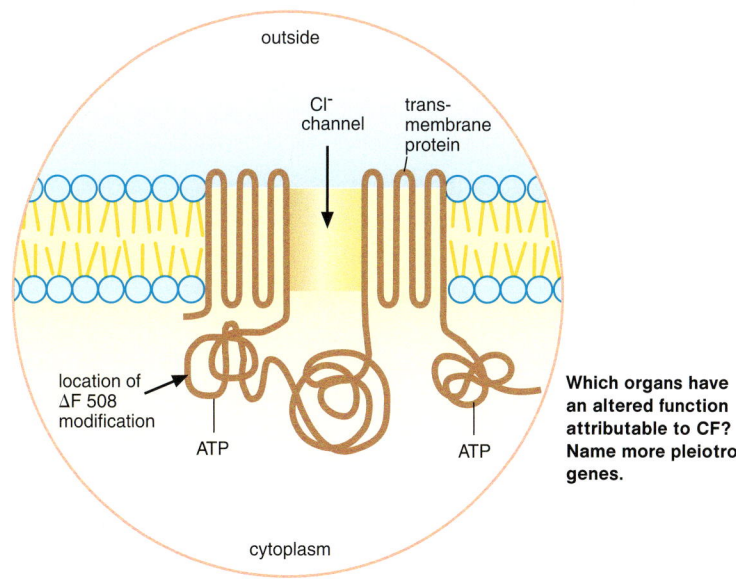

Which organs have an altered function attributable to CF? Name more pleiotropic genes.

The chloride channel

Chloride channels provide the required fluidity of the mucus secreted in the respiratory tract and in the intestine. The chloride channels on the surface of the mucosa actively secrete Cl⁻ ions that attract water osmotically into the mucus. Thereby, the mucus is kept fluid. If this process is disturbed, the thick mucus blocks respiration and causes higher susceptibility to inflammation of the respiratory tract and lungs. The sweat glands reabsorb chloride through these channels. If they are faulty, the reabsorption of chloride ions ceases and NaCl forms. This causes the sweat of CF patients to be significantly more salty than that of healthy individuals.

Somatic gene therapy

Since the CFTR gene has been isolated, an interesting attempt has been made to help CF patients. By using viruses or liposomes carrying this gene, the mucosal membranes in the noses of patients have been experimentally infected, with the restoration of 20 % of metabolic activity. However, the effect only lasts for a few days or weeks. Hence, the therapy has to be carried out several times per year. The aim is to modify those cells of the mucosa that are still able to divide and produce new mucosa cells. These rare cell types (*adult stem cells*) have to be stimulated to incorporate the intact gene.

Gather information regarding the conditions under which gene therapy is allowed in Germany.

Are there hereditary diseases other than CF for which trials of somatic gene therapy are running? Please investigate.

Genetic counselling

Up to 80 % of human embryos with severe genetic defects are aborted in the first few days of pregnancy. Such short pregnancies are often not recognized by these women. Approximately 1–2 % of all live births have a genetic disorder or disability. For this reason, many expectant parents wish to know of any genetic risks to their offspring. This task is carried out by *genetic counsellors*. They also provide information about methods of prenatal diagnosis and possible therapies. The counselling interview prepares the way for any possible result and no decision for or against the pregnancy is demanded.

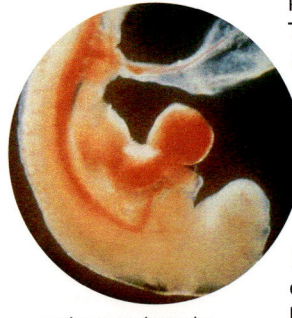

embryo at 4 weeks

Even before a pregnancy, the risks can be estimated by examining the family tree. If a hereditary disease has occurred in the family, the pattern of inheritance can provide information about the risks. In a dominant case, healthy parents have healthy children. If one parent is ill and heterozygous, then statistically 50 % of the children will also be ill. If both parents are heterozygous, this rises to 75 % according to the 2nd Mendelian law. In the case of recessive inheritance, however, the analysis is more complicated, because heterozygous parents appear healthy and can pass the allele for the disease on to their children. In many hereditary diseases, carriers can be identified by a heterozygote test: carriers of sickle cell anaemia have altered blood cells and cystic fibrosis carriers have an increased level of chloride ions in their sweat.

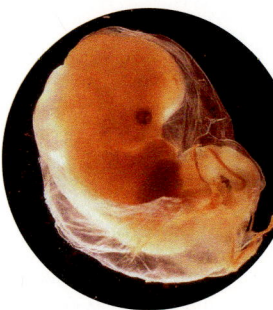

embryo at 8 weeks

Prenatal diagnosis involves medical examinations that can detect possible diseases of the unborn child during pregnancy. This includes *ultra sound examinations* in which the size and possible deformities of the embryo are examined. Thus, not only genetic disorders, but also damage caused by the environment can be diagnosed. Not every *congenital disease* has a genetic cause: virus infections, radiation exposure, metabolic diseases, medication and drug abuse by the mother can harm the unborn child. Here, selective counselling can improve the chances of having a healthy child.

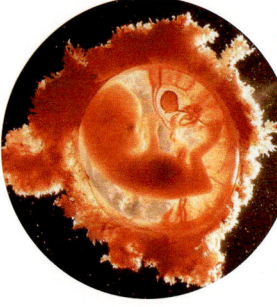

embryo at 14 weeks

Prenatal diagnosis also includes medical interventions in order to obtain cells from the unborn child, namely from the placenta, from the amniotic fluid or from the umbilical cord (*chorionic villus sampling, amniocentesis, umbilical cord vein blood sample,* see page 71). Chromosome damage can be detected by examining prepared cells with a microscope, whereas metabolic diseases are detected by using biochemical tests. Everyone has defective alleles in his/her genome but these have very different outcomes for each carrier and his/her relatives. The risk of suffering from diabetes 2 in old age cannot be put on the same level as the risk of having *Marfan syndrome* (see page 64). Parents are often unaware that particular genetic disorders can be treated. The consequences of *phenylketonuria* can be avoided by means of a special diet (see page 64). Cleft lip, cleft jaw and cleft palate can be corrected by operation and gonosomal anomalies such as *Turner syndrome* (see page 30) can have a milder phenotype following hormone therapy.

Genetic counselling can help parents to make family planning decisions according to their ethical and religious values. If the parents decide in favour of having a child, they can obtain assistance to prepare for life with a potentially disabled child.

Before 1995, a suspected disorder in the child was reason enough to carry out an abortion free of punishment (*indication rule*) by law. In a revised form of §218 StGb of 1995, abortion carried out by a physician remains non-punishable in the first 12 weeks of pregnancy if the pregnant woman can provide proof of counselling with regard to a conflict pregnancy (*period rule*). If, after this period, medical reasons occur that call for an abortion, the period is prolonged until 22 weeks.

Extras

Prenatal diagnosis

Based on the disease and family history of the future parents, the risk of having a child with a hereditary disease can be estimated.

Tasks

① Draw a family tree of the following family and calculate the following probabilities:
  a) The first child of phenotypically healthy parents suffers from an autosomally recessive hereditary disease. What is the probability of the next child also suffering from this disease?
  b) One parent is a carrier of an autosomal recessive hereditary disease, whereas the other parent is homozygous and healthy. What is the probability of the next child being ill? What is the probability that the child is a carrier?

Methods of prenatal diagnosis

When examining amniotic fluid, a needle is inserted through the mother's abdominal wall and into the amniotic sac. Using ultra sound as a guide, the physician extracts amniotic fluid containing foetal cells. Biochemical analysis of the amniotic fluid can reveal metabolic disorders in the foetus.

Foetal cells are obtained by centrifugation of the amniotic fluid and stimulated to proliferate in culture medium. The plant toxin colchicine interrupts mitosis so that easily visible metaphase chromosomes accumulate. These are fixed, stained and examined by light microscopy.
Cells obtained by chorionic villus sampling or umbilical cord vein blood samples are treated in the same way.

Tasks

② Compare the prenatal diagnostic methods concerning the risks to the unborn child and the time that is left for the parents to decide to continue with the pregnancy.

③ Which parents would you advise to use prenatal diagnosis?

④ The *triple test* measures the concentration of three hormones in the blood of the mother (after the 15th week of pregnancy). If the age of the pregnancy is known exactly, the measured values can be used to calculate whether there is an increased risk for Down's syndrome or a developmental defect of the neural tube ("*split spine*", *spina bifida*) in the unborn child. However, many false positive and some false negative results can occur. Discuss the advantages and disadvantages of this test.

Preimplantation genetic diagnosis

The *preimplantation genetic diagnosis* (PID) is a special case of prenatal diagnosis and is part of reproductive medicine. After an in vitro fertilization (see page 60), one cell is removed from the embryo at the 8-cell stage and is examined for genetic disorders. Only after this analysis is it decided whether the embryo should be implanted into the uterus of the woman.

Task

⑤ PGD is prohibited in Germany. List the arguments of supporters and opponents and express your own opinion.

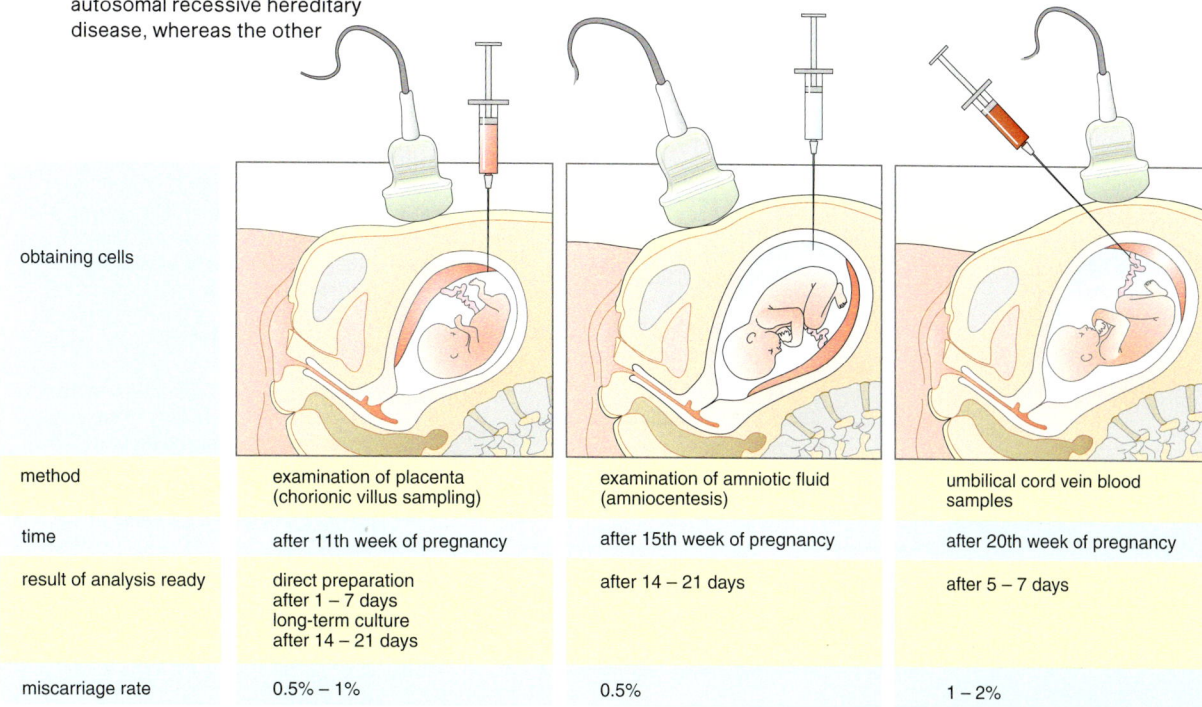

obtaining cells			
method	examination of placenta (chorionic villus sampling)	examination of amniotic fluid (amniocentesis)	umbilical cord vein blood samples
time	after 11th week of pregnancy	after 15th week of pregnancy	after 20th week of pregnancy
result of analysis ready	direct preparation after 1 – 7 days long-term culture after 14 – 21 days	after 14 – 21 days	after 5 – 7 days
miscarriage rate	0.5% – 1%	0.5%	1 – 2%

DNA analysis

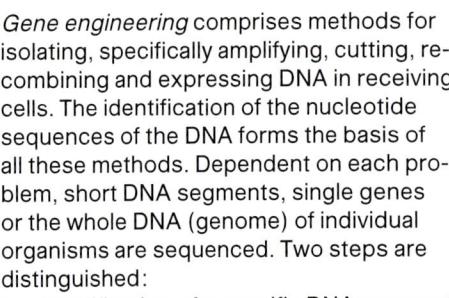

thermocycler

The thermal block for collecting the PCR tubes is on the top of the apparatus. For a PCR, only small amounts of liquid are required, e.g. 1 μl Taq polymerase, 2 μl of each primer 1 and 2.

Gene engineering comprises methods for isolating, specifically amplifying, cutting, recombining and expressing DNA in receiving cells. The identification of the nucleotide sequences of the DNA forms the basis of all these methods. Dependent on each problem, short DNA segments, single genes or the whole DNA (genome) of individual organisms are sequenced. Two steps are distinguished:
1. Amplification of a specific DNA segment.
2. Sequencing of the amplified segment.

Amplification of DNA

The *polymerase chain reaction (PCR)* described by KARY B. MULLIS in 1983 is a method used specifically to amplify DNA segments in a test tube. He solved a problem that had greatly limited molecular biology reasearch. MULLIS used the existing basic knowledge of DNA replication and combined it in a brilliant way (see info box). With the help of PCR, extremely small amounts of DNA samples can be amplified identically in a short period of time. PCR has revolutionized basic genetic research and application-oriented disciplines (e. g. medical diagnostics, criminalistics).

DNA sequencing

DNA sequencing is based on a method described in 1977 by FREDERICK SANGER and ALAN COULSON. Today, many variants are based on the same principle to terminate DNA synthesis in a test tube with modified nucleotides.

Suspended DNA polymerase is mixed with *deoxynucleotides* (dNTP) and *dideoxynucleotides* (ddNTP) (see fig. 73. 2). This mixture is added to DNA single strands amplified by PCR (see info box). During the synthesis of complementary DNA strands from DNA single strands, both nucleotide

»info box«

Polymerase chain reaction (PCR)

Polymerase chain reaction (PCR)
A method by which DNA segments are selectively amplified in a test tube

At high temperatures (96 °C, 2 min), double-stranded DNA is melted (= separated into single strands). The labelling of the DNA segment that should be copied is an essential step in every PCR: the nucleotide sequence of the ends of the particular segment must be known. Short *complementary* DNA molecules are

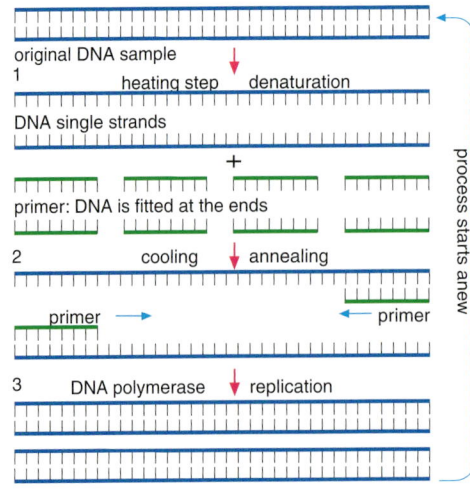

synthesized. They are used as *primers* that "ignite" DNA synthesis by the DNA polymerase. In order to set up replication, a pair of primers each running in opposite directions is used. At 50 °C (1.5 min), the primers attach to the DNA templates. Starting at the 3' end of the primers, DNA polymerase synthesizes the complementary strand. *Taq polymerase* is used for this DNA synthesis (at 70 °C). This enzyme was first isolated from the bacterium *Thermus aquaticus*, which likes heat and lives in hot springs. It has an exceptionally high temperature optimum at 72 °C and survives the high melting temperatures during the PCR.

Primers and *nucleotide triphosphates* from which the new DNA strands are made are added in excess in the beginning of a PCR.

A *thermocycler* (see margin) is used for PCR. The core piece of these apparatus is a thermal block that can change its temperature. The number and duration of the cycles and also the duration of each of the three single PCR steps are programmed and proceed automatically.

types are incorporated. If dNTP is incorporated, the synthesis continues. However, this
is not the case with ddNTP because it lacks
the OH group at the C3 atom to which the
polymerase usually adds the next nucleotide. The early termination of DNA synthesis
thus occurs in some newly made strands.
If sorted by length, the synthesized fragments differ by one nucleotide (fig. 1). In the
laboratory, the Sanger-Coulson samples are
then separated by electrophoresis in a sequencing gel. Smaller strands migrate faster
in the direct current than longer strands. In
non-radioactive methods, dd-nucleotides
are coloured differently with fluorescence
dyes (fig. 2). Thus, the DNA fragments have
these "markers" on their 3' ends. Fragments
having the same size form a band and pass
a detector on their way through the gel; the
detector detects the colour order. By reading the colour order, the base sequences
can be identified (fig. 3).

Now that Taq polymerase, a temperature
resistant DNA polymerase, is available, the
Singer-Coulson method can be performed
like a PCR and contains the following three
steps: denaturation, marking with a primer
and elongation. Several cycles are performed to enhance the amount of the products. When compared with PCR, there is
an essential difference: only one of the two
primers can be used, so that only one of the
two template single strands is replicated.
The previously completed PCR completed
is therefore carried out either in such a way
that the primers are completely used up or
the template strand used for sequencing
has to be isolated.

Tasks

① A PCR cycle in a test tube resembles
    DNA replication in cells. Explain.
② The newly synthesized DNA strands
    are used as templates for the following
    cycle. Calculate the number of strands
    after 30 cycles.
③ Complete the base sequence (fig. 3).
④ What is the base sequence of the tem-
    plate DNA strand?

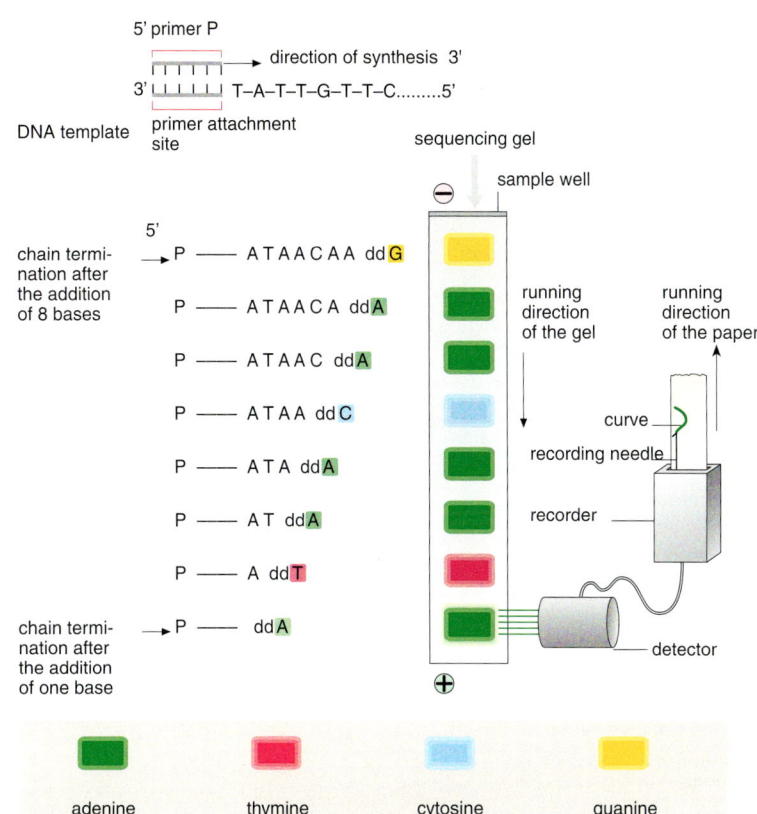

1 Representation of DNA sequencing by SANGER-COULSON

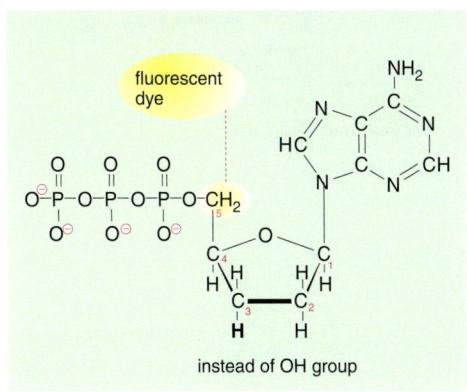

2 Dideoxyadenosine triphosphate (ddATP)

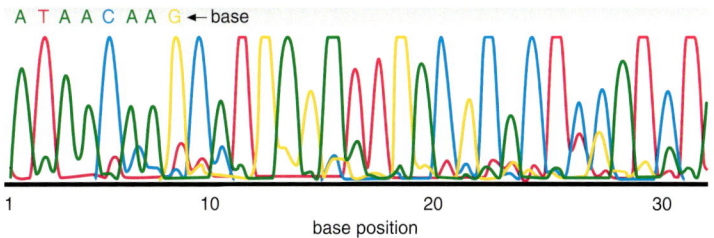

3 Printout of the fluorescence detector

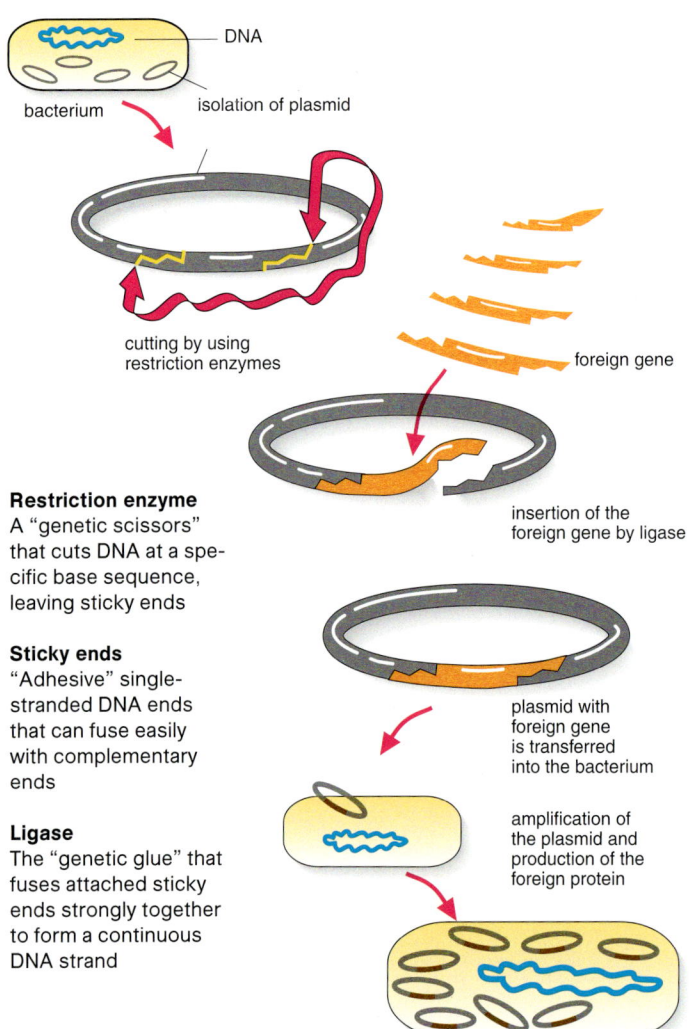

Restriction enzyme
A "genetic scissors" that cuts DNA at a specific base sequence, leaving sticky ends

Sticky ends
"Adhesive" single-stranded DNA ends that can fuse easily with complementary ends

Ligase
The "genetic glue" that fuses attached sticky ends strongly together to form a continuous DNA strand

DNA

bacterium

isolation of plasmid

cutting by using restriction enzymes

foreign gene

insertion of the foreign gene by ligase

plasmid with foreign gene is transferred into the bacterium

amplification of the plasmid and production of the foreign protein

1 Production of a genetically modified bacterium

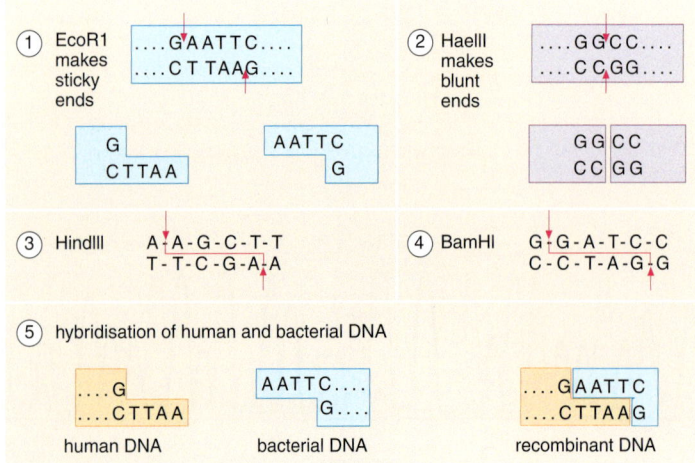

2 Effect of various restriction enzymes

Biotechnology

Archaeologists have found out that yeast cells were used for beer production in Great Britain, as early as about 4,000 years ago, at the end of the New Stone Age. The use of microorganisms by humans is therefore not a new invention of modern biotechnology. In the 20th century, however, such usage had reached industrial scales. At first, only microorganisms could be cultivated and then only for the production of antibiotics (e. g. penicillin) and other active substances, namely compounds that were natural parts of these organisms. Drugs such as insulin are only produced by higher organisms and not by microorganisms. The rapid advance of biotechnology during the past few decades has been based on our ability to isolate the gene of an important protein and to incorporate it into another organism. The organisms that have a genome that has been changed in such a way are called *transgenic organisms*.

Genetic scissors and glue

In 1972, WERNER ARBER isolated bacterial enzymes that could specifically cut DNA. These so-called *restriction enzymes* do not cut randomly but only within specific sequences of about 4 – 5 bases in length. Such a sequence occurs in the DNA about every 1,000 to 10,000 nucleotides. The fragments are of the respective size. Several dozen restriction enzymes are known. Most of them do not leave a "blunt" end.

The restriction enzyme EcoR1 of *Escherichia coli*, for example, cuts at its recognition sequence in every DNA double strand between the bases G and A. Since they are not opposite to each other in the double strand, the cut produces overhanging ends whose base sequence is complementary to each other. They attract each other and are therefore called *"sticky ends"*. If the DNA of different species is cut with the same restriction enzyme, it is possible to fuse the fragments at their complementary ends. The fusion is stable when the DNA is linked into a continuous strand. This task is carried out by an enzyme called a *ligase*.

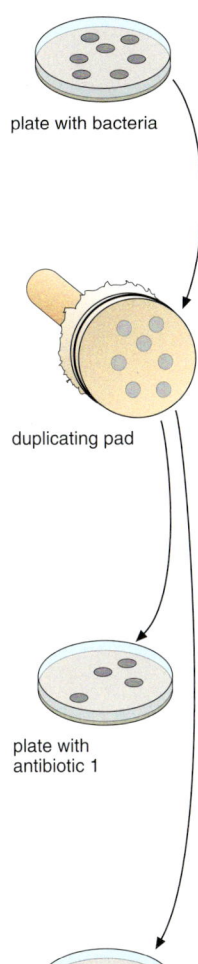

plate with bacteria

duplicating pad

plate with
antibiotic 1

plate with
antibiotic 2

Gene transfer

If we wish to insert and express a gene, e. g. for human insulin, in the bacterium *E. coli*, a *vector* is needed as the gene vehicle. Bacterial *plasmids* are often used for this purpose. Plasmids are circular DNA molecules found abundantly in the cytoplasm together with the "bacterial chromosome". Plasmids have the ability to replicate (see page 46) and they often carry genes that provide the bacterium with resistance against particular antibiotics. Bacteria can take up plasmids from the culture medium via a process called *transformation* (see page 46). Bacterial plasmids that contain two antibiotic resistance genes and the insulin gene are cut with the same restriction enzyme. They are mixed in a test tube so that they fuse. Then, the constructs are transferred into a suitable *E. coli* strain.

On culture medium containing antibiotic 1, only those bacteria grow and form colonies that have taken up the plasmid with the resistance gene. In order to identify the bacteria that have taken up a plasmid with the insulin gene, a trick is used: the employed plasmids have an additional resistance gene against antibiotic 2. The insulin gene is incorporated into the middle of this resistance gene, thereby disabling it. Thus, only bacteria that survive in the presence of antibiotic 1 but not antibiotic 2 contain the plasmid with the insulin gene.

Extraction of the gene product

In order to obtain the protein encoded by the foreign gene, transgenic bacteria are cultivated in suitable conditions. Since not only the gene segment coding for the foreign protein, but also the associated promotor has been transferred, the foreign protein can be transcribed, translated and finally isolated from the bacterial cultures.

Tasks

1. Eukaryotic genes are often interrupted, i. e. they have coding parts (*exons*) separated by non-coding parts (*introns*). Prokaryotic genes have no introns. What difficulties can arise when eukaryotic DNA is transferred into prokaryotic bacteria?

2. The enzyme reverse transcriptase has the ability to "re-translate" mRNA to DNA. Imagine you obtain mature human insulin mRNA from the pancreas and the enzyme reverse transcriptase. How would you proceed experimentally to solve the problem in task 1? Plan an experiment.

3. After their translation on the ribosomes, many eukaryotic proteins are brought to the endoplasmic reticulum and modified there, e. g. carbohydrate residues are often attached. In bacteria, post-transcriptional modification does not exist. Additionally, no disulphide bridges are formed between the residues of the amino acid cysteine in proteins such as insulin. What problems might result from insulin production in bacteria?

4. Foreign genes have been introduced, by laboratories, into many plant species of agricultural importance. This was performed to set up resistance against viruses, herbicides and drought. Most plant vectors carry a gene with resistance against the antibiotic kanamycin. Animal tests caused concerns that the product of the resistance genes might be toxic for humans. Why do plant vectors contain resistance genes? What medical or ecological problems might arise if these resistance genes are present in food?

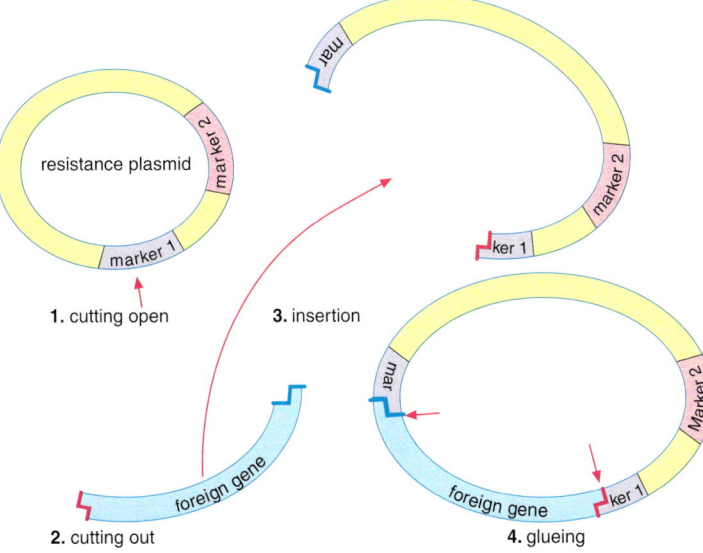

1. cutting open
2. cutting out
3. insertion
4. glueing

resistance plasmid

marker 2
marker 1
marker 2
ker 1
mar
foreign gene
foreign gene
ker 1
Marker 2
mar

1 Construction of an expression vector

Genetic markers and hybridization probes

Headlines such as "Suspect convicted thanks to a saliva sample!" are no longer rare. Today, DNA analysis methods play an important role in criminology. Biological material (blood residues, sperm, hair or saliva on a cigarette) is as reliable as a forgery-safe identification card. Such identification is also referred to as *genetic fingerprinting*. A few cells are enough for analysis because the sample material can be amplified by PCR (see page 72). The parts of the human genome that code for proteins are not analysed. These DNA segments do not differ greatly from person to person. Mutations in these parts are often selection-neutral because they have no effect on the phenotype. Therefore, they have accumulated during evolution. In present-day criminology, usually SNPs (see page 67) and microsatellites (see info box page 77) are analysed.

DNA fragmentation

To analysis RFLPs (*restriction fragment length polymorphisms*), the sample DNA is cut with suitable restriction enzymes and analysed. Restriction fragments of different sizes are based on differences in the base sequences of the cutting sites so that the enzyme can no longer cut the DNA. The loss of a cutting site shows up in gel electrophoresis when two fragments stay linked together (see fig. 1).

The restriction samples are separated by using electrophoresis. For the subsequent process, single-stranded DNA is needed. Therefore, the electrophoresis gels are treated with an alkaline solution to denaturate the DNA. The restriction fragments are then transferred onto a nylon membrane (*blotting*, see page 9), on which they stick, for further analysis. Radioactively marked probes are added. Probes are short artificially produced DNA single strands that are complementary to the bases of selected non-variable sites of the restriction fragments. This, however, requires the knowledge of these base sequences. The probe molecules bind (*hybridize*) to the complementary bases of the restriction fragments. In order to make them visible, the nylon membranes are placed onto radioactive-sensitive film. The pattern of the restriction fragments (each fragment of specific length forms a band) can be seen as blackened

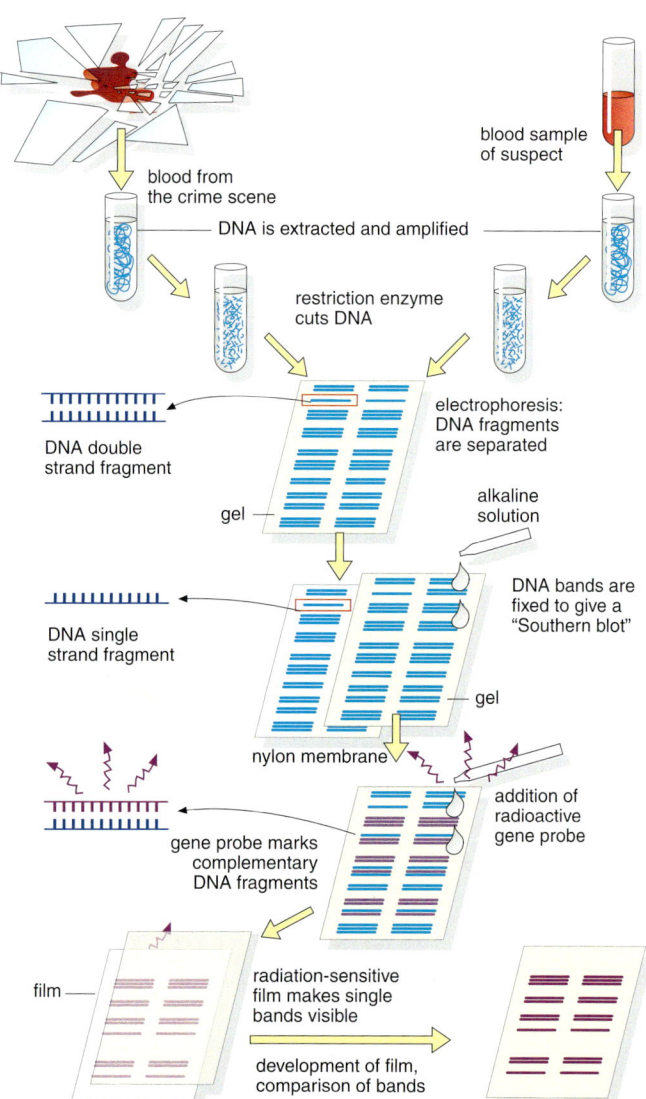

1 Production of a genetic fingerprint (RFLP analysis)

2 Genetic fingerprinting from a paternity test

band on the film. Figure 76. 2 shows the restriction fragment pattern of various people. Depending on their degree of relationship, related people show similar patterns. Restriction fragments are inherited based on the Mendelian laws.

Today, this method is used mainly in human genetics. If fatherhood (paternity) is in doubt, DNA from the mother, the child and the man in question are analysed. Since the child has half of its genetic material from its mother and the other half from the father, a comparison of the restriction fragment patterns shows whether the man is really the father. If bands occur in the child's pattern that are neither found in the pattern of the mother nor of the man in question, he can be eliminated as the biological father.

If genes have to be detected in chromosomes (see info box) or mRNA has to be detected in tissue sections, fluorescently marked probes are used (see picture in info box). The marked probes are added to the prepared chromosomes or tissue sections. They attach to the complementary DNA or mRNA segment. The location of the probe can be detected based on the fluorescence signal seen via a fluorescence microscope.

Tasks

① In figure 76. 2, comparable sections of the genetic fingerprints from a mother (M), her daughter (D) and possible fathers (F_{1-4}) are shown. Which man could be the biological father? Explain.

② A specific gene ("breast cancer gene") mutation is regularly found in breast cancer patients. A causal connection is however uncertain because some women have this mutation but no breast cancer. Imagine that a woman has had her DNA tested and that the result is positive. What problems might she now face? Keep in mind that there is as yet no effective therapy.

≫info box≪

Gene mapping

Since crossing experiments cannot be performed with humans, genes are mapped by using family tree analysis. In order to find out whether the risk of having breast cancer is hereditary, DNA samples from three generations of affected families were taken and used for restriction fragment length polymorphism analysis.

A specific DNA fragment was seen in all affected women. This fragment was sequenced and, by using probes, was shown to be located on chromosome 17. The region considered for the potential breast cancer gene contained 20 mega bases and about 1,000 genes. The exact mapping was successful by using small base segments (microsatellites) made up of highly repetitive base sequences (e. g. CACACA). *Microsatellites* are about 200 nucleotides long and repeat themselves up to 20 times. The human genome has about 650,000 such microsatellites and they serve as genetic markers. Microsatellites were found that were located close to the potential breast cancer gene and were inherited with it. Thus, the region in which the gene must be located was reduced down to 600 kilo bases, sufficient room for 60 genes.

For further analysis, the finding was used that breast cancer often occurs together with ovary cancer. The mRNA of both tissues was prepared from patients and translated back to DNA by using reverse transcriptase. DNA fragments of both tissues bound only to the 600-kb region in the genome of the patients. The synthetic base sequence therefore corresponded to the chromosomal sequence. Thus, the gene responsible for breast cancer has probably been identified. Its exact function, however, remains unknown.

Fluorescently marked DNA probes bind to complementary chromosome segments. If the sequence of the chromosome segment is known, the probe can be added for attachment and the incorporated probes can be seen via a fluorescent microscope when the probes are excited by light of an appropriate wavelength (see fig., yellow regions). This method can also be used for diagnosis.

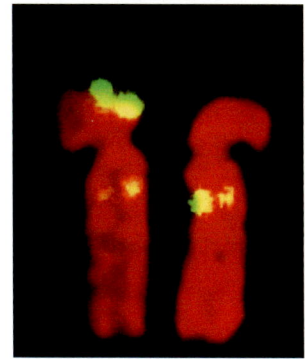

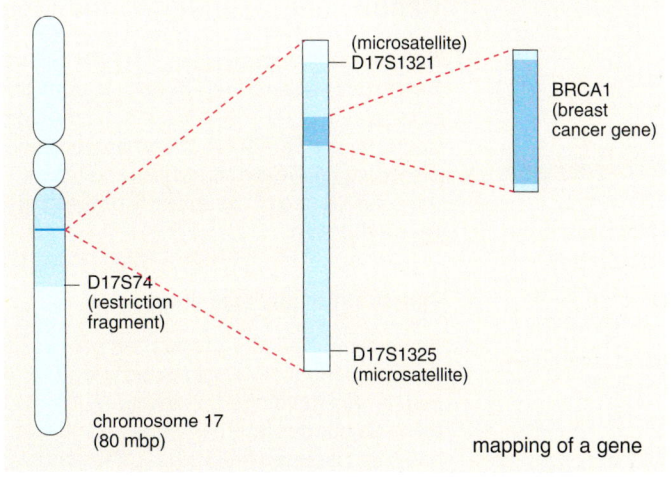

mapping of a gene

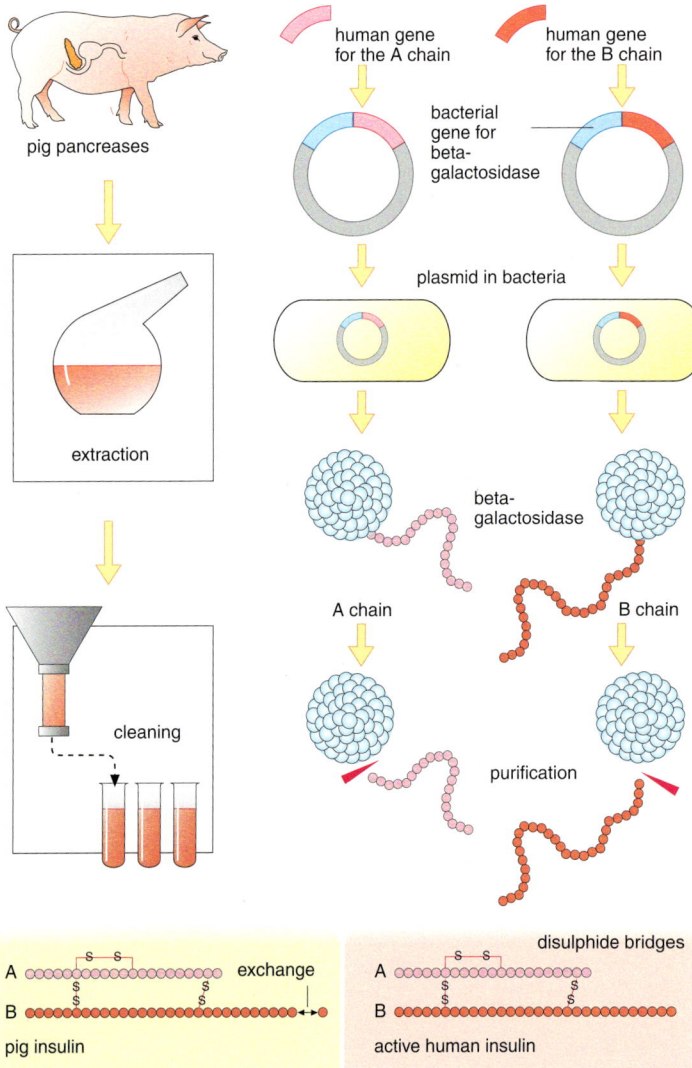

A ●●●●●●●●●●●●●●●●	exchange	disulphide bridges
B ●●●●●●●●●●●●●●●●●●●●●		A ●●●●●●●●●●●●●●●●
pig insulin		B ●●●●●●●●●●●●●●●●●●●●●
		active human insulin

1 Classic and gene engineering methods of insulin production

Genetic engineering in medicine

Genetic engineering has profoundly changed the diagnostic and therapeutic possibilities of medicine. This concerns the production of drugs, the detection of certain genetic disorders (*genetic diagnosis*) and also the possibilities of the therapy of diseases (*gene therapy*).

Genetically engineered drugs

As early as 1982, the production process for the first genetically engineered drug *human insulin* was developed. Previously, insulin that was needed to treat diabetes mellitus had been isolated from pigs' pancreas. Insulin is quite a short protein with an A chain and a B chain linked by two disulphide bonds (fig. 1). By controlled linking of single nucleotides, both corresponding DNA sequences can be artificially made and each one can be transferred into bacteria by a plasmid. There, they are controlled by same promoter as beta-galactosidase. The bacteria are optimally supplied with nutrients and oxygen in enormous *fermenters*. Modified (*transgenic*) bacteria now produce the two peptide chains of human insulin linked to beta-galactosidase. As soon as enough proteins have been made, the cells are broken open and the peptide chains are separated from beta-galactosidase and linked to each other by two disulphide bonds forming the end product. Insulin production was improved when it became possible to synthesize one DNA sequence for both insulin chains. The produced prohormone forms the two disulphide bonds independently. Meanwhile, a range of drugs can be produced by transgenic bacteria or yeast cultures (see page 83).

Genetic diagnosis

If we know the base sequence of genes that cause hereditary diseases, we can detect them by using complementary probes. Such genes can then be detected in people not yet suffering from the disease. The carrier can be informed, for example, about methods of prevention to lower the possible risk of getting the disease. *Huntington's chorea* is a dominant hereditary disease that only manifests itself after the age of 25 to 55 and causes the loss of brain tissue. This results in paralysis, the inability to carry out movements and the slow loss of brain activity. The disease is based on the amplification of a DNA segment and is as yet uncurable. A negative test result can bring great relief to people who have a family history of the disease. However, how might people deal with a positive result and the certainty of becoming ill?

In contrast, the identification of the breast cancer gene only means that the carrier has an elevated risk of developing breast cancer. Additionally, only every 20th breast cancer is based on this gene defect (see page 77). In most diseases, genetic diagnosis only provides a risk estimation. Even DNA chip technology will not be able to change this, despite its efficiency (see info box).

Examples of drugs produced by gene engineering

1982: human insulin against diabetes

1983: factor VIII against haemophilia

1985: interferons against leukaemia

1986: vaccine against hepatitis B

1987: growth hormones

Somatic gene therapy

Somatic gene therapy aims at establishing the intact version of the gene causing the disease in the somatic cells of the patients. In order to treat *cystic fibrosis* (see page 68), DNA segments are enclosed in, for example, liposomes and inhaled by the patients. These segments might be taken up by the epithelial cells of the lung but are only expressed for a few weeks. Researchers have tried to transfer *tumour suppressor* genes to cancer patients or to make cancer genes ineffective. Often cells are taken from the patient, are genetically modified and then transferred back to the patient.

Germ line gene therapy

In *germline therapy*, an intact gene is transfered into a fertilized egg cell by microinjection. The zygote is then implanted into the uterus. The gene is later present in all cells of the growing individual and will also be expressed. In humans, it is theoretically possible to influence hereditary diseases. However, other inherited features might be unintentionally changed and the risks are incalculable. Therefore, germ line therapy is prohibited in Germany and in many other countries: genetic changes to gametes and the cells developing from them are not allowed.

Gene pharming

Genetic modifications of germ line cells are not prohibited in animals. In the so-called *pharming* method (wordplay on "farming" and "pharmaceutical"), such transgenic animals are used for the production of drugs. The DNA coding for the desired product is transferred into the nucleus of a mammal zygote in such a way that it is only expressed in the udder of the adult animal. The gene product can then be isolated from the milk. For example, factor VIII is currently produced in this way to treat haemophilia (see page 63).

»info box«

DNA chips

Tens of thousands of single-stranded DNA molecules can be placed onto a chip of glass or silicon. These synthesised DNA chips might provide an easy and cost-effective way of testing DNA sequences of gene samples in the future.

All gene chips work on the same principle: in gene-specific chips, the surface is divided into a dot matrix. Each dot contains a defined variant of a single-stranded DNA molecule. The examined DNA sample is separated into single strands and labelled with a fluorescent dye and added to the chip. The DNA of the sample that pairs with the exactly complementary DNA sequence on the dot matrix leaves a signal (see fig.). Knowing the sequence on the dot matrix, the sample DNA can be identified. If the sample DNA gives a signal, for example, in a segment containing a mutant, the sample must contain exactly this defect. The chip signals are analysed by a scanner connected to a computer.

On the other hand, DNA can also be isolated from patients, amplified by PCR (see page 76) and attached to a surface. If fluorescently labelled DNA molecules of known sequence are now added (e.g. a gene for a specific disease), this single-stranded probe will only pair with the appropriate complementary DNA. A signal means that the patient is a carrier of this gene.

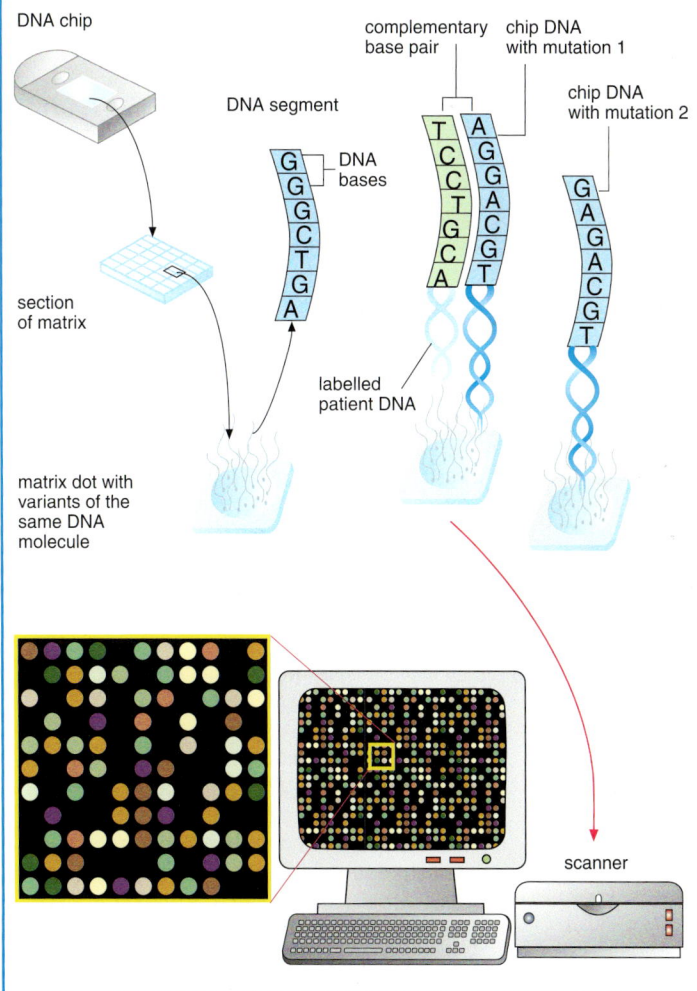

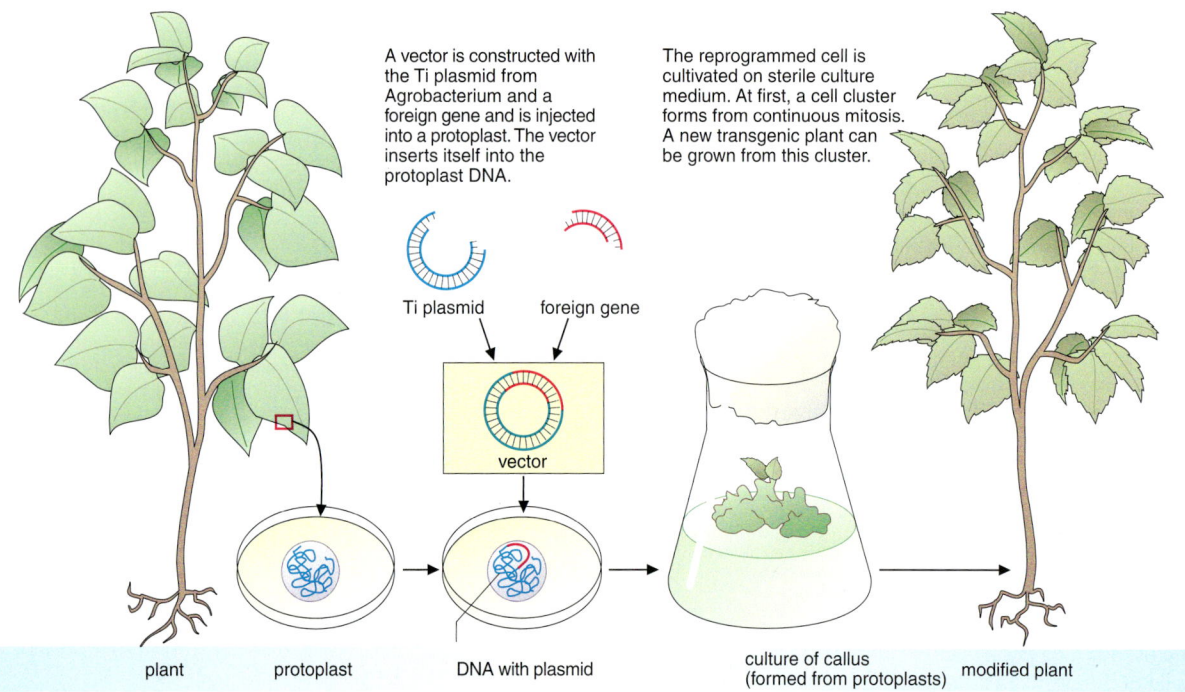

A vector is constructed with the Ti plasmid from Agrobacterium and a foreign gene and is injected into a protoplast. The vector inserts itself into the protoplast DNA.

The reprogrammed cell is cultivated on sterile culture medium. At first, a cell cluster forms from continuous mitosis. A new transgenic plant can be grown from this cluster.

Ti plasmid   foreign gene

vector

plant   protoplast   DNA with plasmid   culture of callus (formed from protoplasts)   modified plant

1 Transfer of a foreign gene to a plant (Ti plasmid technique)

Genetic engineering in agriculture

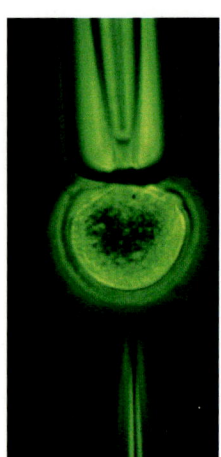

Genes are "transplanted". DNA is injected into a plant cell by a micropipette.

Gene engineering in agriculture aims to transfer desired features to crops or farm animals or to turn off undesired features. In contrast to common breeding, species differences are no longer important: it is possible to transfer genes from bacteria, fungi, animals and humans. *Green gene technology* works with transgenic plants and *red gene technology* with transgenic animals (e. g. gene pharming, page 79).

For the transfer of genes, the bacterium *Agrobacterium tumefaciens* has been shown to be efficient. It causes crown gall disease, a plant tumour. It enters the plant via injuries and then transfers its tumour-inducing plasmid (*Ti plasmid*) into single cells. The plasmid becomes integrated into the plant DNA and causes uncontrolled growth in the plant. A plant tumour forms and excretes nutrients for the bacteria. The bacterium's ability to transfer DNA is being used in gene engineering: if the tumour-inducing base sequences are removed and a foreign gene is incorporated into the plasmid, the desired genes can be integrated into the plant DNA. If a marker is also transferred at the same time (e. g. an anti-biotic resistance gene), the success of the gene transfer can be tested. The transfer is easiest to carry out in young plant cells whose cell wall has been removed by enzymes. Under suitable culture conditions, these protoplasts can be stimulated to divide and whole plants can be regenerated. During this asexual reproduction, no recombination of genetic information takes place. Thus, the transgenic feature cannot be lost in the process.

In 2001, 54 *transgenic plants* were licensed worldwide. So far, soy beans, cotton, maize and oilseed rape have the greatest economic importance. Often, this method is used to achieve increased plant resistance against insects, fungi, bacteria and viruses. Frequently, herbicide resistance genes (71 %) are transferred that make the plant non-susceptible to specific weed killers. Opponents of genetic engineering question this method, as the companies that produce the transgenic plants also sell the respective herbicides, thereby strongly inducing farmers to buy only their products.

Tomatoes and other soft fruits are often picked when they are still green to avoid them being squashed before reaching the supermarkets. Unfortunately, these unripe fruits frequently lack full flavour.

The expression of several genes is responsible for the ripening of a tomato. These genes make a tomato red, aromatic, soft and finally squishy. The "squishy" gene is important for the degradation of cell walls. A non-squishy tomato was the first genetically engineered plant that was allowed on the market. In it, the "squishy" gene is turned off by an *antisense method*. The "squishy" gene is integrated inversely after the same promoter. If it is transcribed, an mRNA complementary to the original is produced. Normal (*sense*) and *antisense mRNA* attach and no translation occurs (fig. 1). The enzyme degrading the cell wall is thus no longer produced. The transgenic "anti-squishy tomatoes" ripen slower and the red fruits remain firm for 14 days longer than ordinary tomatoes. Therefore, they can be picked and dispatched when they are ripe and aromatic. However, the tomato ages in an invisible way and its vitamin content goes down just as in a common tomato. From 1994, this tomato has been licensed in the USA and, from 1996, the selling of tomato paste made from such tomatoes has been allowed in the EU (see info box).

In 1999, "vitamin A rice" was introduced onto the market by Swiss researchers. Its yellow colour (*"golden rice"*) is attributable to its content of beta-carotene, the precursor of vitamin A. Golden rice was developed in two steps by transferring seven different genes to the rice genome. Apart from beta-carotene, it contains other substances that improve the availability of iron for the human body. This rice sort might help to improve nutrient conditions in developing countries. However, it has been shown that a person must eat about 7 kg to cover his daily need of vitamin A.

Task

① Gather information regarding arguments for and against genetic engineering using the examples herbicide resistance gene, insect resistance and genetically engineered tomatoes.

»info box«

The novel food regulation of the EU from 1997

Genetically modified food or food that is produced by genetically modified organisms (GMO) is classified as "novel food":

Category 1: GMO or food containing material from GMO, the genetic modification of which is detectable. This food has to be labelled as *"genetically modified"* (e. g. herbicide-resistant soy beans).

Category 2: Products derived directly from foods in category 1 and in which the genetic modification is detectable (DNA or the respective protein) also have to be labelled (e. g. soy protein).

Category 3: Nutrients that are produced by GMO or by enzymes derived from GMO but which cannot be biologically distinguished from common foods (no labelling requirement), e. g. soy oil.

Category 4: Products that are a risk to the health or environment cannot be approved for the free market (e. g. allergy-causing foods).

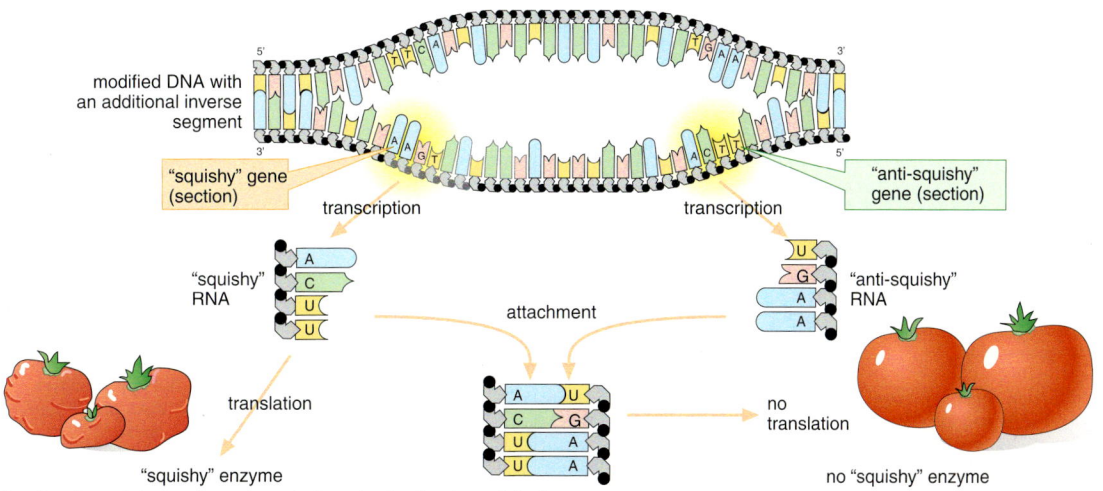

1 Application of the antisense technique in the "anti-squishy" tomato.

Gene ethics

Gene menu

Tomatoes that look red and feel firm, even after weeks of storage.

Potatoes that are waxy and have their own built-in pesticide.

Meat of the pink-pigmented "giant" pig.

Giant carp refined with a trout gene.

Hard cheese ripened in only a few days by "Maxiren" (Maxiren is produced by a genetically-modified strain of *Kluyveromyces lactis* and is used as a vegetarian replacement for calf chymosin during the coagulation of milk proteins in cheese making)

Delicate "killer" cheese that kills its own mould.

Beer genetically engineered to be "light" by using super yeasts.

Sweetener made from fresh intestinal bacteria.

Enjoy your meal!

Gather information about the foods in the above gene menu and if they can be offered for sale in Germany according to current laws.

To the question: "Would you eat foods containing genes?", 70 % of the questioned people answered "no". Half of all Germans believe: "Vegetables usually do not contain any genes."

Carry out a survey and demonstrate the results in a presentation.

Obtain information about the opportunities provided by and the risks of genetically modified foods and set up discussion groups that are pro and contra genetically modified food.

Nomen est omen?

The risks and opportunities presented by genetic engineering are often controversially discussed by expert groups and in public. Frequently, we can detect on which side the groups stand by the terms that they use:

— genetically changed/optimised/manipulated
— research on genetic safety/risk
— mixing/addition/impurity/contamination of seeds with genetically modified components

Look for more such terms and explain their psychological effects.

Better plants?

"What are we going to do if the continuous change in climate requires new plants with improved features?

Do we then have the time to cross our agricultural crops blindly until perhaps new and better adapted plants arise?" (GEN SUISSE: Foundation for Responsible Gene Engineering)

"So far the experiences gained from cultivating genetically manipulated plants in Europe and other regions show that genetically engineered seeds are already out of control. The efforts to control gene engineering fail miserably once it is set free in the environment. In the

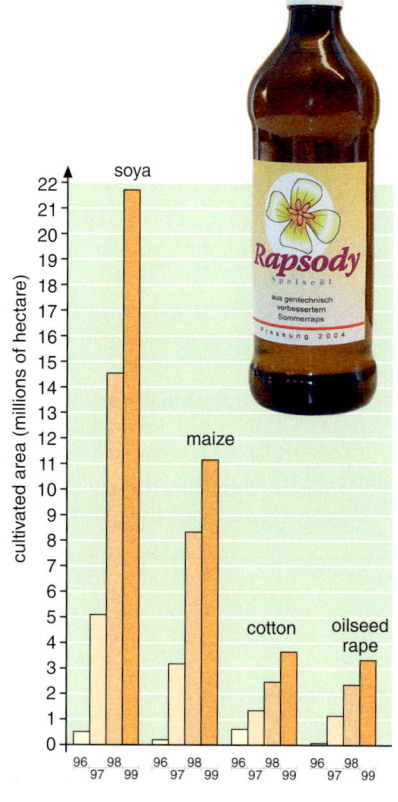

light of these dangerous deficiencies, the responsibility of giving new approval cannot be taken."

(Greenpeace, Germany)

Search for information about the cultivation of genetically modified plants in Germany. Discuss the opportunities provided by and the risks of the cultivation of transgenic plants.

Who decides?

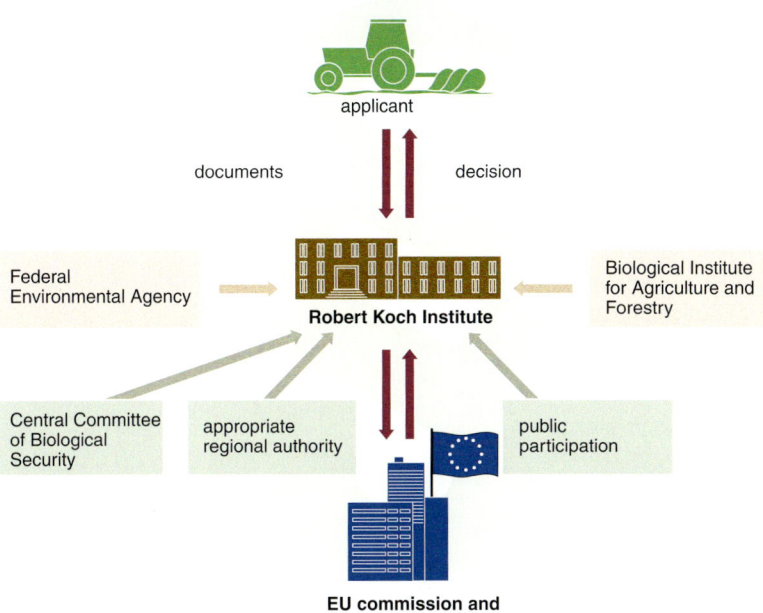

applicant

documents → ← decision

Federal Environmental Agency → Robert Koch Institute ← Biological Institute for Agriculture and Forestry

Central Committee of Biological Security | appropriate regional authority | public participation

EU commission and EU member states

---

ment or "designer babies". It's only a matter of time, many say, before parents will improve their children's intelligence and personality by having suitable genes inserted into them shortly after conception." (Guardian 2003)

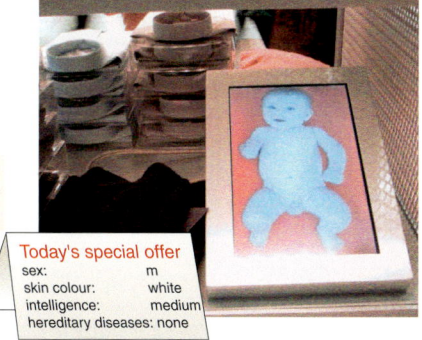

Today's special offer
sex: m
skin colour: white
intelligence: medium
hereditary diseases: none

> **The combination of gene engineering and reproductive technology contains possibilities that produce hope and fear. Collect arguments in order to make the discussion more objective.**

To whom do genes belong?

"Genetic resources are the common heritage of mankind."
(Bonner Guidelines 2002)

Whole branches of industries are currently searching for usable genetic characteristics in organisms. About 86 % of the plants that are used by healers in Samoa have been shown to have biological activity in laboratory tests. The knowledge of such healers can lower the research expenses of the pharmaceutical companies significantly.

Biodiversity convention (Nairobi, 2000): "The fair distribution of the advantages of the application of traditional knowledge, innovations and customs of many indigenous and local communities with traditional life style is desirable."

> **The search for genes, whose genetically engineered expression could be used commercially, is called "bioprospecting" by supporters and "biopiracy" by opponents. Explain.**

Custom-made humans?

"In the year 2003, the first European designer baby was born in England. Stem cells of the baby's umbilical cord should be able to help its 4-year-old brother who has cancer. Only absolutely identical stem cells can be used and, for this reason, the embryo was actively selected during in vitro fertilization." (agency announcement)

"This year's 50th anniversary of the discovery of the structure of the DNA has kindled many debates about the implications of that knowledge for the human condition. Arguably the most emotionally charged is the debate over the prospect of human genetic enhance-

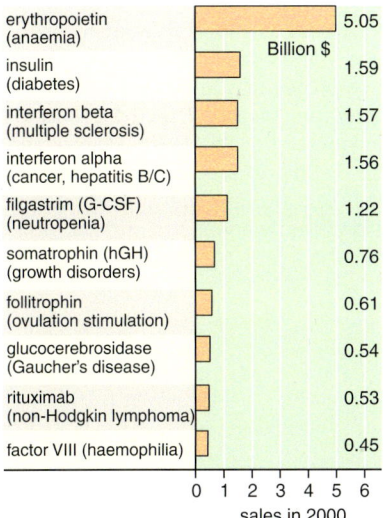

	Billion $
erythropoietin (anaemia)	5.05
insulin (diabetes)	1.59
interferon beta (multiple sclerosis)	1.57
interferon alpha (cancer, hepatitis B/C)	1.56
filgastrim (G-CSF) (neutropenia)	1.22
somatrophin (hGH) (growth disorders)	0.76
follitrophin (ovulation stimulation)	0.61
glucocerebrosidase (Gaucher's disease)	0.54
rituximab (non-Hodgkin lymphoma)	0.53
factor VIII (haemophilia)	0.45

sales in 2000

Data protection for genes?

The HUGO Ethics Committee adopts the following principles as a basis for its recommendations:

1. Human genomic databases are global public goods.
2. Individuals, families, communities, commercial entities, institutions and governments should foster the public good.
3. The free flow of data and the fair and equitable distribution of benefits from research using databases should be encouraged.
4. The choices and privacy of individuals, families and communities with respect to the use of their data should be respected.
5. Individuals, families and communities should be protected from discrimination and stigmatization.
6. Researchers, institutions and commercial entities have a right to a fair return for intellectual and financial contributions to databases.

> **Gather information regarding the way that a genetic profile of individuals could be used for bone marrow donations, insurance purposes, crime investigation, choice of employee and immigration and the manner in which it is already being used in Europe.**
>
> **Develop ideas for effective data protection of personal genetic information.**

Recognition and defence

Countless microorganisms (bacteria, fungi, protozoa) and viruses are present in our environment. These *infectious agents* or *pathogens* are transmitted by physical contact, tiny droplets, air and food, but also by insect bites. They can cause great damage in the body if they multiply. Usually, in healthy people, most infections only last for a short time and do not cause serious health problems. The body's defence systems are responsible for this.

Our body is protected against infections by several mechanisms. Surface barriers (*external defence*) effectively inhibit the entrance of agents into the body. The skin, for example, is slightly acidic (*acid mantle*) and most bacteria cannot multiply there very well. Mucous membranes excrete secretions in which bacteria are enclosed. In the bronchi, foreign material is transported to the outside by cilia. Tear fluid contains *lysozyme*, an enzyme that destroys the cell wall of many bacteria.

If the infectious agents overcome the outer barriers, e. g. after burns or skin injuries, the immune system becomes activated. It is specialized to recognise and destroy foreign invaders. The ability to distinguish the body's own and foreign material is the basis of any internal defence system.

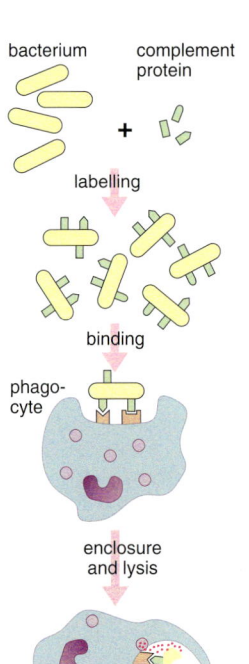

lymphatic organs

bacterium   complement protein

+

labelling

binding

phago-cyte

enclosure and lysis

lyso-some

Mode of action of the complement system

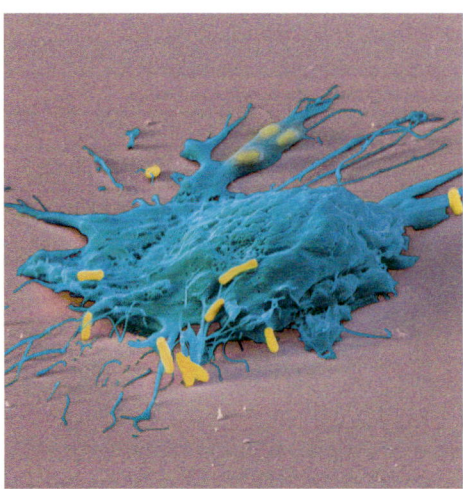

1 A macrophage (blue) seeks bacteria (yellow)

Innate immunity

Macrophages occur mainly in the skin and belong to a group of scavenging cells (*phagocytes*) that attach to the surface of foreign material (fig. 1). The foreign material is then absorbed into the cells by enclosing it into small membranous vesicles.

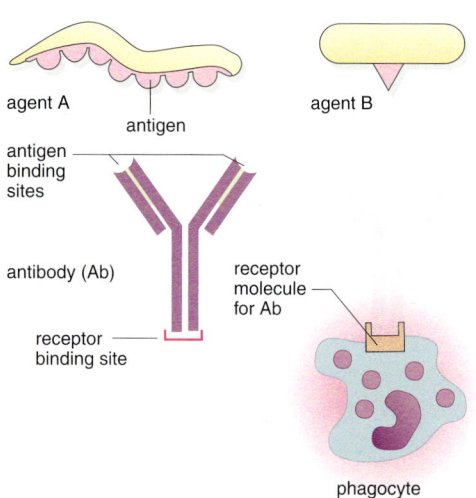

agent A   agent B
antigen
antigen binding sites
antibody (Ab)   receptor molecule for Ab
receptor binding site
phagocyte

2 Antibody bridges: agent — phagocyte

Lysosomes fuse with these vesicles. Lysosomal digestive enzymes degrade the foreign material. The entire process (attachment, adsorption, digestion) is called *phagocytosis*.

Phagocytosis is enhanced by specific proteins found in the body fluids of mammals. All of these proteins together are called the *complement system*. They adhere to foreign invaders and label them. Phagocytes recognize these proteins by means of their receptor molecules (see margin). Phagocytosis is significantly enhanced and the multiplication of the agent is inhibited.

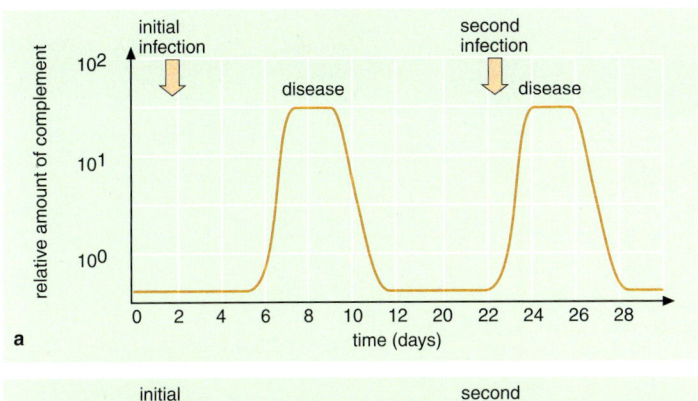

a

b

1 Comparison of innate (a) and adaptive immunity (b)

Inflammation

If agents penetrate our skin and multiply, easily visible, red and swollen tissue areas often occur. The organism has responded to the infection by increasing the blood supply to the infected area and by increasing the permeability of the capillaries. Millions of immune cells leave the blood vessels, enter the infected tissue and migrate to the source of infection (fig. 2). In combination with the highly increased concentration of complement proteins and antibodies, the infection can be fought successfully. _Pus_ mainly contains immune cells that have died after phagocytosis.

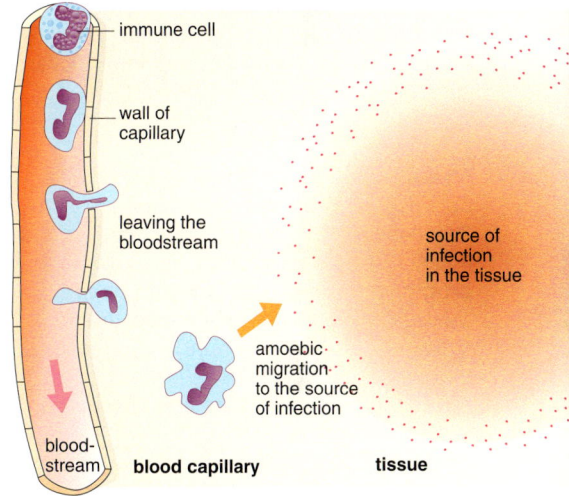

2 Immune cells migrate to the source of infection

Passive immunization
Antibodies specific for the antigen are given

Active immunization
Weakened agents or parts of the agent are given — production of antibodies by white blood cells

Incubation period
Time interval between the agent's invasion of the body and the appearance of the first disease symptoms

Adaptive (= acquired) immunity

Only some agents that intrude into the body are labelled by the proteins of the complement system. The rest evade this control. More specific adapter molecules are needed to fight these agents. They should ideally be able to bind to all imaginable agents and additionally contain binding sites for the receptor molecules on phagocytes. _Antibodies_ have these abilities (see fig. 84.2). The adaptive immune system of mammals can produce antibodies against almost all surface molecules found on agents. Substances that cause antibody production are called _antigens_. Antibodies are produced by specialized white blood cells. The adaptive immune system does not return to its initial state after successfully fighting an infection. It has a "memory" and _memory cells_ remain: if the same agent invades the body again, the memory cells multiply and produce a higher concentration of antibodies and in a shorter time than during the first infection _(secondary immune response)_. The antibody-labelled agents cannot multiply and are quickly destroyed. The body is immune against this agent.

Tasks

① Cells and adapter molecules of the immune system have to work together to fight an infection successfully. Explain.

② What is the difference between innate and adaptive immunity in mammals? Relate your answer to fig. 1.

③ Vaccination (active immunization) is based on immunological memory. Explain this by using fig. 1.

The immune response

Phagocytes represent the phylogenetically oldest defence system that is found in all groups of animals today. During evolution, vertebrates have additionally acquired a *lymphatic system* as a response to constantly occurring new infectious agents. The lymphatic system (see margin) is composed of the lymphatic organs (e. g. thymus and lymph nodes), lymph vessels and various white blood cells. The vessels of the lymphatic system take up fluid from the surrounding tissue and drain it into the bloodstream through the *thoracic duct*. The fluid flows through the lymph nodes in which white blood cells are present at high concentrations. Antigens in the tissue fluid are recognized by these cells, which bind and destroy them. Lymph nodes can be considered as immunological filters. If there is an infection, they usually become swollen.

The origin of white blood cells

All cells of the human immune system are derived from stem cells in the bone marrow. Two *lineages* are distinguished, the *myeloid* and *lymphoid lineages* (see fig. 1). Myeloid progenitors can differentiate into *phagocytes* (monocytes, macrophages, granulocytes) or *mast cells*. Lymphoid progenitors become *lymphocytes* (B cells and T cells). All the mentioned cells are able to migrate and can leave the bone marrow actively to enter the blood. They are flushed away by the bloodstream and can leave the blood vessels in all tissues. In the tissues, some of the monocytes become macrophages. It is estimated that hundreds of billions of these phagocytes constantly patrol our tissues in order to detect foreign invaders. B cells are located especially in the lymph nodes. If they come into contact with an antigen, they produce antibodies and become *plasma cells*. T cells leave the bone marrow in an immature state. They migrate to the *thymus*, a gland-like structure behind the sternum (breast bone). There they become mature *helper T cells* and *cytotoxic T cells*. Mature T cells have antigen-specific binding molecules on their surface. After maturation, T cells leave the thymus and populate almost all tissues in the body.

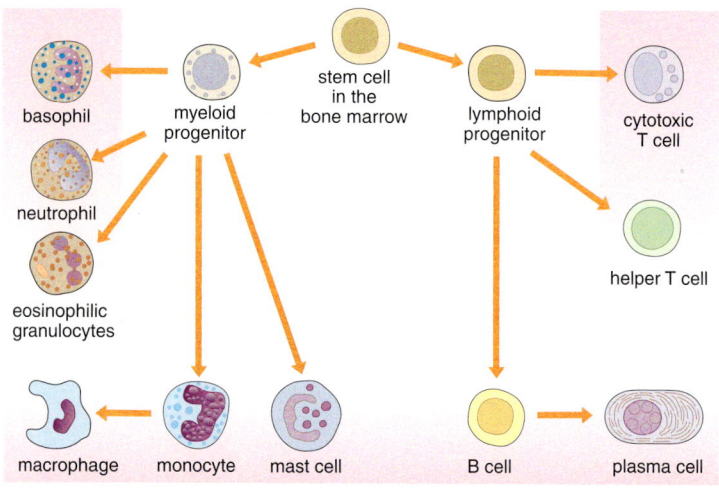

1 Differentiation of white blood cells

Antibody production

If bacteria penetrate our body, they are phagocytosed and digested by macrophages. Specific bacterial protein fragments are transported to the surface of the phagocytes (see fig. 87. 1). These fragments thus become a detectable antigen for helper T cells (T_H). Helper T cells have a key role during antibody production because they provide the essential help required by the B cells to become antibody-producing plasma cells. T_H cells recognize the fragments presented by the macrophages with their T cell receptor and then become activated. Using its membrane-bound antibodies, a B cell that has attached an antigen presents fragments of it in a similar way as the macrophages do. If an activated T cell meets such a B cell, it stimulates it hormonally. Only now is the B cell able to divide, with its daughter cells becoming plasma cells that produce suitable antibodies. The antibodies are released into all body fluids. The cooperation between macrophages, T_H cells and B cells assures that the produced antibodies fit exactly to the agent that has invaded the body. The antibodies bind to this agent; the phagocytes recognize the labelled invader and eliminate it by phagocytosis. The activated T and B cells die after successfully fighting the infection.

Lymphocytes
B cells: mature in mammals in the bone marrow; in birds in a thymus-like bursa fabricii.
T cells: mature in the thymus

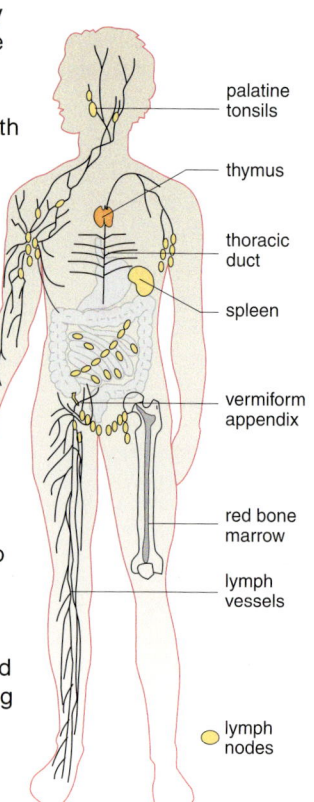

palatine tonsils

thymus

thoracic duct

spleen

vermiform appendix

red bone marrow

lymph vessels

lymph nodes

lymphatic system

Elimination of virus-infected cells

The so-called *cytotoxic T cells* (T_C) are specialised in the elimination of virus-infected cells. T_C cells constantly patrol our body. If they meet a virus-infected cell in the tissue, they bind to it and release molecules that kill the infected cell. T_C cells use their T-cell receptor to recognize infected cells by fragments that they present on their surface (fig. 2).

Immunological memory

If an agent enters the body fluids for the first time, B and T cells are activated by the mechanisms described above. The antigen-specific cells multiply and some of the T_H and plasma cells differentiate into defending cells, with others becoming long-living *memory cells*. These memory cells are immunologically inactive at first. However, if the agent enters the body again, the memory cells are activated within a very short period of time. They divide rapidly and, in a short time, they produce large amounts of antibodies (secondary immunological response). The agent is fought efficiently so that no or only a few symptoms of the disease occur.

Tasks

1 T-cell receptor molecules are of great importance for antibody production and the killing of virus-infected cells. T-cell receptor molecules recognize fragments of foreign invaders on antigen-presenting cells by the *lock-and-key principle*. Which cells present antigens on their surface?

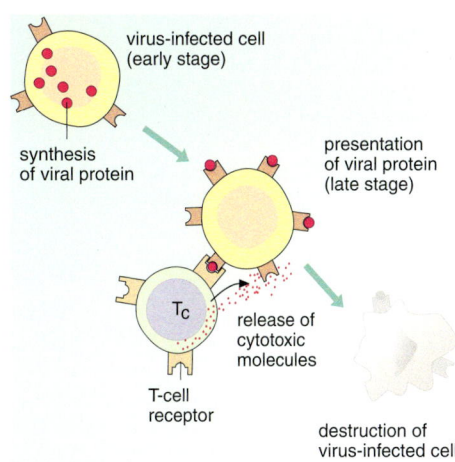

2 Cytotoxic T cell eliminates a virus-infected cell

2 A single B cell is only specific for one single antigen; this means that its antibodies will bind only to this particular antigen. What mechanism assures that activated helper T cells only help such B cells that can produce the correct antibody?

3 Helper T cells pass to the lymph nodes via the lymph fluid after their activation in the tissues. B cells occur in high concentrations in the lymph nodes. This ensures that helper T cells have a high probability of meeting B cells presenting the correct fragment. B cells can therefore remain inside the lymph node and still carry out their function as plasma cells in contrast to cytotoxic T cells. Explain this phenomenon.

4 How do cytotoxic T cells distinguish virus-infected cells from normal cells?

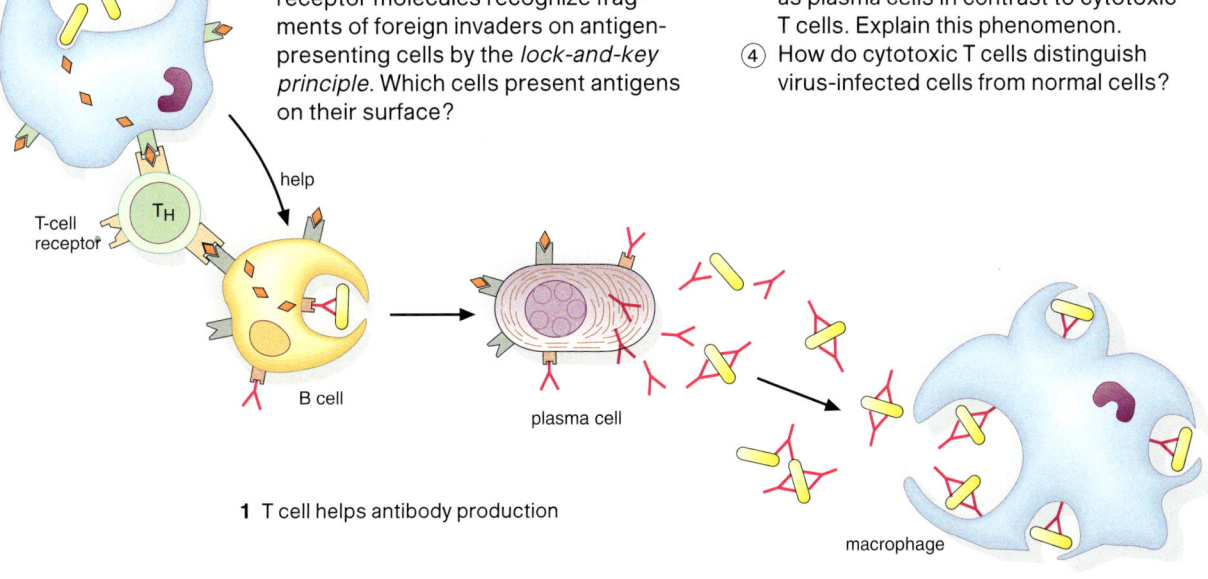

1 T cell helps antibody production

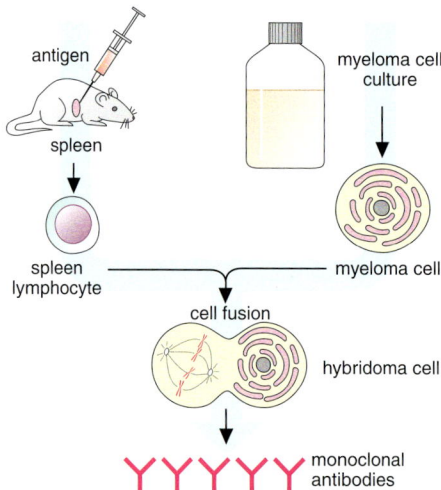

1 Production of monoclonal antibodies (mAb)

Monoclonal antibodies

The ability of antibodies to bind to defined epitopes is used in many ways, e. g. to identify the *BSE pathogen* or tumour cells. In research, this method is used to demonstrate cell structures or to identify cell types by using surface molecules called *cell markers*.

In order to produce monoclonal antibodies, laboratory animals (e. g. mice) are immunized with the purified antigen of another species (e. g. human). The antibody-producing lymphocytes (more precisely plasma cells) are extracted from the spleen of the mice. In a test tube, they are then fused with cancer cells by using various chemicals. The fusion products are called *hybridoma cells*. They still produce the specific antibodies of the plasma cell and are furthermore able to divide in an unlimited fashion like cancer cells. Basically, they can be kept in culture forever. In order to identify the hybridoma clone that produces the desired antibodies, the fused hybridoma cells are separated into single cells and then cultured separately.

Antigen-antibody reaction

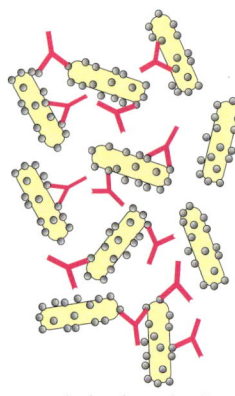

antibody binds to epitope

Antibodies are present in the blood and tissue fluids of all vertebrates. All antibodies have the same basic structure. They are Y-shaped molecules consisting of two identical heavy and two identical light amino acids chains (see margin). The chains are connected by disulphide bonds. The molecule also possesses a constant region that is identical within an antibody class (see page 90). This is distinguished from the variable region, which is responsible for antigen binding. Based on the lock-and-key principle, the antigen-binding sites recognize and bind specific structures of an antigen called its *epitope* (see margin). Since each antibody molecule has two identical antigen-binding sites, they can link foreign invaders together. The agglutinated foreign material is eliminated by phagocytosis.

agglutination of cells by antibodies

When identifying blood groups, it is usually sufficient to obtain sera from various people. They contain a mixture of various antibody specificities including some that specifically recognize substances of the blood groups in the ABO system. They then link and agglutinate the red blood cells carrying that specific antigen. In order to identify cell markers, e. g. to distinguish helper from cytotoxic T cells, these sera usually do not suffice. The antibodies in the sera have many specificities for various cell surface molecules and do not distinguish between these antigens clearly enough. Here, antibodies are needed that recognize only one defined epitope: *monoclonal antibodies* (mAb).

»**info box**«

ELISA

For an ELISA (Enzyme-Linked ImmunoSorbent Assay), the purified antigen used for immunisation is affixed to the surface of the sample plate. The *monoclonal antibodies* to be tested are added to each test tube.

If the corresponding antibody is present, it binds to the antigen fixed to the surface. The tubes are washed to remove unbound antibodies. A secondary antibody is added that binds to the monoclonal antibody.

The secondary antibody is linked to an enzyme that converts a colourless substrate to a colourful stain. Hybridoma clones that contain the desired monoclonal antibody (mAb) can be detected by the production of colour.

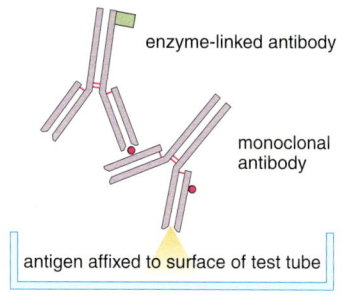

Clone
Group of genetically identical cells or genetically identical organisms

The antibodies that are secreted by these *cell clones* into the culture supernatant are purified and tested by means of an ELISA test for their specificity (see info box page 88).

Cell-type analysis using mAb

The complicated processes occurring during the interaction of various cells during an immune response (see page 86) could only be clarified after it was possible to distinguish the involved cell types in a test tube by using mAb. Human T cells have various surface molecules (fig. 1). Helper and cytotoxic T cells have a set of common surface molecules that are not suitable for discrimination. The surface molecules CD4 and CD8 enable identification because they are each only present on one cell type. Anti-CD4 and anti-CD8 mAb are marked with different fluorescent pigments. The marked mAb are added to the examined cells and, under the fluorescent microscope, the cytotoxic and helper T cells glow with different colours if they bind mAb.

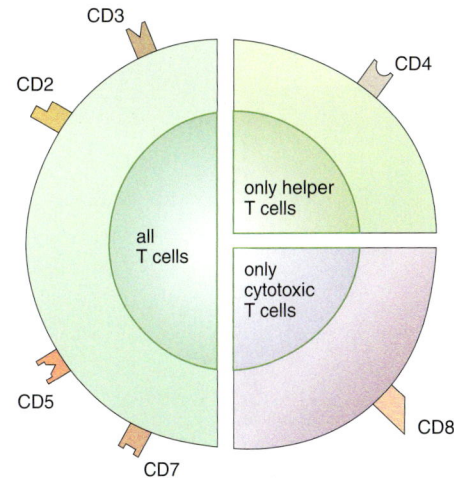

1 Surface molecules of T cells

»info box«

Antibodies produced according to the building block principle

The immune system of vertebrates is able to produce millions of different antigen receptor molecules (antibodies, T-cell receptors). If every single antibody was represented by a DNA segment, it would use up the largest part of the genome. Antigen receptor molecule diversity is achieved by the rearrangement of a manageable number of DNA segments during the maturation of B and T cells.

The DNA segment that contains information about the production of the light (L) antibody chain is made up of numerous gene segments (V, J, C) that are interrupted by non-coding regions. The V and J segments code for the antigen-binding variable part, and the C segment codes for the constant part. V segments are present in 200 variations. During the maturation of a B cell, one V segment is selected by chance and added to the J segments (*somatic recombination*).

All remaining V segments in the DNA are removed by *deletion*. The pre-mRNA of a mature plasma cell is a copy of this rearranged segment. Introns and all J segments except one are removed by splicing (see page 23). The mature mRNA is translated, on ribosomes, into the L chain. This mechanism enables the production of $200 \cdot 4 = 800$ different L chains. H chains are produced by the same principle but additionally one of 12 different D segments are included after the V segment. There are then $200 \cdot 12 \cdot 4 = 9600$ different H chains possible. Since the H and L chains pair form a complete antibody, there are $800 \cdot 9,600 = 7.68$ million possible combinations.

Susumu Tonegawa (Nobel Prize 1987) discovered the reason for the great diversity of possible antibody molecules despite the relatively small number of corresponding genes.

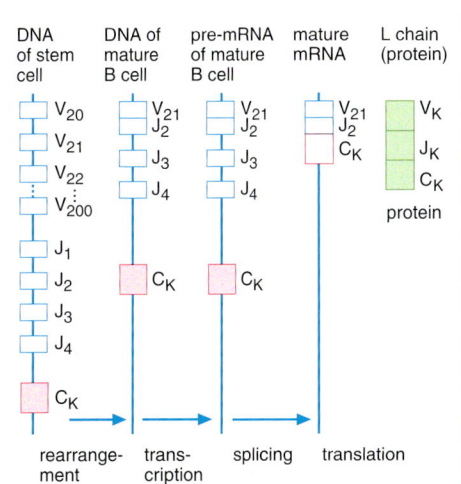

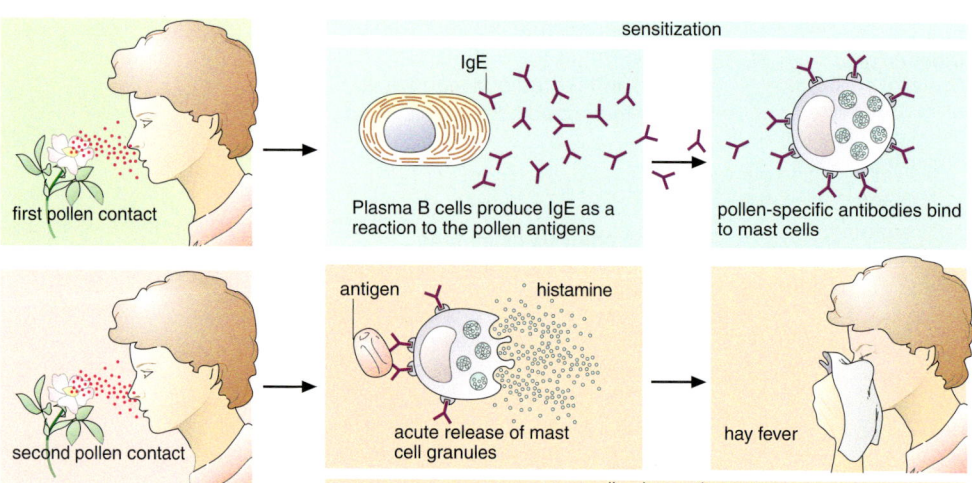

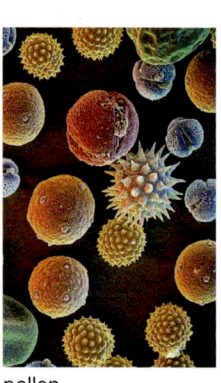

pollen

1 Processes occurring during allergy of the immediate type

Within figure 1:
first pollen contact

Plasma B cells produce IgE as a reaction to the pollen antigens

IgE

sensitization

pollen-specific antibodies bind to mast cells

second pollen contact

antigen

histamine

acute release of mast cell granules

hay fever

allergic reaction

Adverse immune reactions

The price that an organism pays for having a complex immune response is the risk of hypersensitivity or false reactions of the immune system. Whereas antigen-antibody reactions usually go on without being noticed and protect the organism, they can also have harmful effects.

Allergies

A typical hypersensitivity reaction is an *allergy*. This manifests itself in many people, for example, as *hay fever*: especially in the spring and summer when the air is full of pollen, affected people react by a reddening of the eyes and sneezing. Allergy-causing antigens such as pollen are called *allergens*. Hay fever and allergic bronchial asthma are allergies of the *immediate type*. This means that, after the first contact with the allergen (*sensitization*), the next contact with the same allergen will cause a strong reaction straight away. Plasma cells react against the antigen by producing mainly antibodies of the IgE class. The Y-shaped molecules attach with their constant end to membrane-bound receptors on mast cells. These IgE-loaded mast cells are activated as soon as an antibody binds to the allergen. They release *histamine* and other highly reactive substances into the extracellular space. Within minutes, this leads to blood vessel dilation, oedema and nettle rash (hives) (fig. 1). In hay fever, these reactions are usually limited to a specific area. After an insect bite or drug intolerance, however, the reaction can become systemic and spread

to the entire body causing a life-threatening decrease in blood pressure and bronchial cramps (*allergic or anaphylactic shock*).

Many substances such as soaps, cosmetics, synthetic fibre, nickel and chrome can act as contact allergens and cause eczema on the skin. Contact allergies are allergies of the *delayed hypersensitivity reaction type*. They occur days or several weeks after antigen contact and are mediated by cells, especially T lymphocytes and macrophages.

The treatment of allergic reactions is difficult because of their various causes. It is best to avoid allergen contact. If this is not possible, then *desensitization* can often help: regular, continuously increasing doses of allergen cause the release of additional antibodies of the class IgG. These capture the allergens before they can activate the mast cells. In acute reactions, drugs that bind histamine can help immediately.

Transplantation and transfusion

An immune reaction similar to an allergy occurs during the rejection of foreign tissue after transplantation. Here, the cellular immune defence is also activated. In particular, *cytotoxic T cells* and *macrophages* destroy the foreign cells. The foreign tissue is rejected even faster and more strongly if specific proteins on the donor and the recipient cells are different (*transplantation antigens*). Prior to each transplantation, the tissues of the donor and

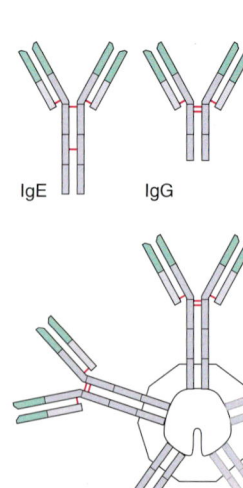

IgE   IgG

IgM

antibody types IgM, IgG and IgE

recipient are tested for compatibility. In most cases, the transplantation antigens do not match perfectly and the immune reaction of the recipient has to be suppressed by drugs (*immune suppression*). However, this can also lead to complications because the patient becomes highly susceptible to infections. In some severe cases, the immune cells of the transplant attack the immunosuppressed host (*graft-versus-host disease*). Today, kidneys, bone marrow, cornea, liver, heart and pancreas can be successfully transplanted.

Compared with tissue transplantation, the success rate of a *blood transfusion* is much higher because blood cells carry fewer transplantation antigens than other cells. Of the several hundreds of *blood group systems*, the *ABO* and the *rhesus system* are of main importance. Without prior contact to foreign blood cells, every human has IgM antibodies in his blood serum against those antigens of the ABO system on the red blood cells that do not occur on his own cells. It is assumed that these antibodies were originally produced against antigens of intestinal bacteria, which have antigens similar to the blood group antigens.

Blood of the blood group A contains antibodies against antigen B (anti-B), whereas blood of the blood group O contains anti-A and anti-B. If a recipient with the blood group A accidentally receives blood of group B, the blood cells aggregate and block the capillaries. This aggregation is an immune reaction of the recipient against antigens of the foreign blood cells. This reaction can also be observed in a test tube (blood group test, see margin).

The *rhesus factor* of blood cells is formed by several antigens (C, D, E), with D being the most effective. Therefore, people with blood cells carrying antigen D are simply called *Rh-positive* and those without are *rh-negative* (d). In contrast to the ABO system, D antibodies develop only after a rh-negative person has come into contact with Rh-positive blood. This sensitization can happen not only during an incompatible blood transfusion, but also during birth or miscarriage if the blood of a Rh-positive child passes into the maternal blood system. A rh-negative mother then produces IgG antibodies against the Rh-positive blood cells of the child. On the second contact, e. g. the next pregnancy with a

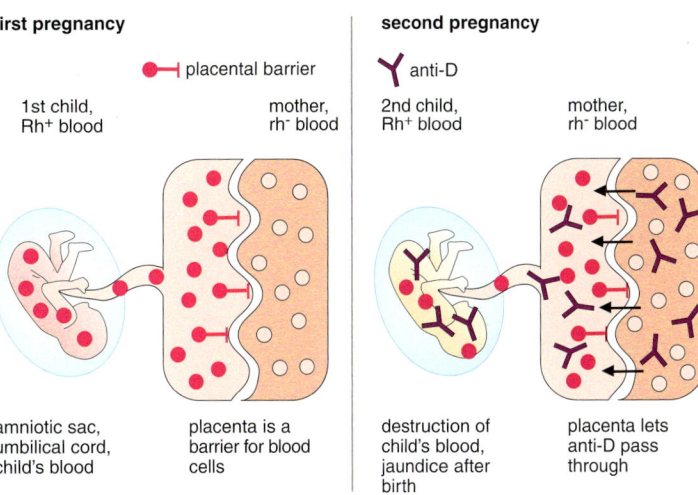

first pregnancy

●⊣ placental barrier

1st child, Rh⁺ blood   mother, rh⁻ blood

amniotic sac, umbilical cord, child's blood   placenta is a barrier for blood cells

second pregnancy

Y anti-D

2nd child, Rh⁺ blood   mother, rh⁻ blood

destruction of child's blood, jaundice after birth   placenta lets anti-D pass through

1 Rhesus incompatibility

Rh-positive child, these antibodies can cross the placenta, reach the foetal blood system, destroy the blood cells and harm the baby. Here only an immediate blood exchange of the newborn can save its life. This disease can be avoided if the mother receives anti-D prophylaxis straight after the birth. For this purpose, anti-D antibodies are injected and mask the children's D blood cells and therefore inhibit sensitization.

Autoimmune diseases

Usually the immune system attacks only foreign cells or substances and spares its own. If this differentiation between "own" and "not own" is defective, antibodies can be produced against the body's own substances or tissues. This causes severe *autoimmune diseases*. About 90 % of patients with *diabetes mellitus* type 1 have antibodies in their blood against their own cells in the islets of Langerhans in the pancreas. This finally cause the cessation of insulin production. Chronic inflammatory rheumatism is also an autoimmune disease. The body produces antibodies against its own IgG molecules and the immunocomplexes thus formed lead to joint inflammation. In almost all patients, these antibodies can be detected (*rheumatoid factor*).

	anti-gens	anti-bodies
A	A	anti-B
B	B	anti-A
AB	A B	none
O	none	anti-B anti-A

blood group features

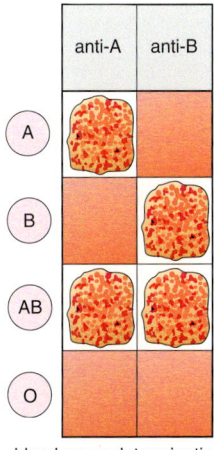

	anti-A	anti-B
A		
B		
AB		
O		

blood group determination by agglutination

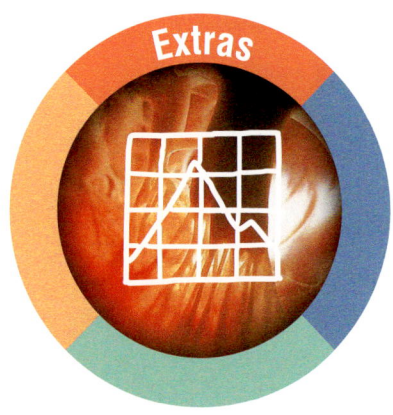

Transplantation antigens

All human cells have on their surface a range of proteins (*transplantation antigens*) that play a central role in the rejection of an organ transplant. If the antigens of the donor and receiver are compatible, the transplanted organ is usually not rejected. The antigens of blood cells can be identified by using antibodies. The corresponding data of potential organ donors and patients are collected in a central database. If a donor dies of brain death, the patient that best qualifies as a receiver of his organs can be quickly determined.

Genetics of the HLA system

The genes coding for human transplantation antigens are located on chromosome 6. The complex of all these genes is described as the *HLA system* (**h**uman **l**eukocyte **a**ntigen). There are 4 gene loci adjacent to each other on chromosome 6 (HLA-A, -B, -C, -D, fig. 1). These genes are linked and inherited together. Since human cells are diploid, an individual can be homozygous or heterozygous for each of these 4 gene loci. The situation becomes even more complicated

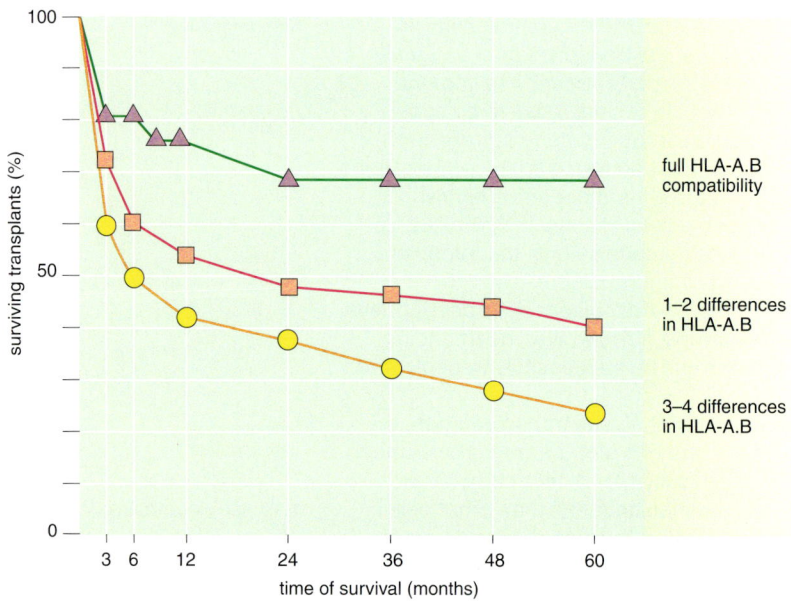

2 Success rate of transplantations

because many HLA alleles exist in the human population, i.e. there are *multiple alleles*.

Figure 3 shows the inherited units (*haplotypes*) of the HLA-A and HLA-B gene loci within one family. The multiple alleles are marked as A1, A2,..., A10, B1, B2,..., B35.

Tasks

(1) Gather information about the legal foundations of organ donation in Germany. Use, for example, the Internet to research this topic.

(2) Use fig. 2 and explain why the success of organ transplantation depends on the degree of compatibility of the HLA system.

(3) Use fig. 3 and show that the genes of the HLA system are linked and inherited together.

Transplant rejection

Transplants are mainly destroyed by cytotoxic T cells and are thus rejected. These cells need helper T cells to become activated (see page 87).

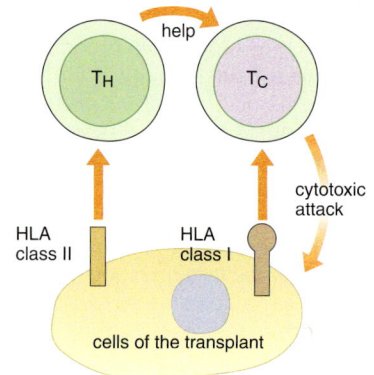

4 Transplant rejection

Tasks

(4) Describe the cooperation of the two cell types during transplant rejection.

(5) HLA antigens are those surface structures that present antigens. The complex is recognized by T-cell receptors. Why is this important for the immune response?

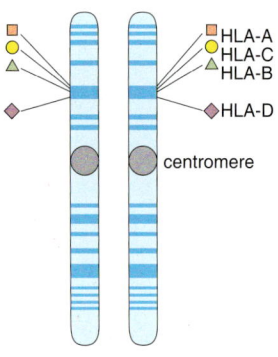

1 Human chromosome 6

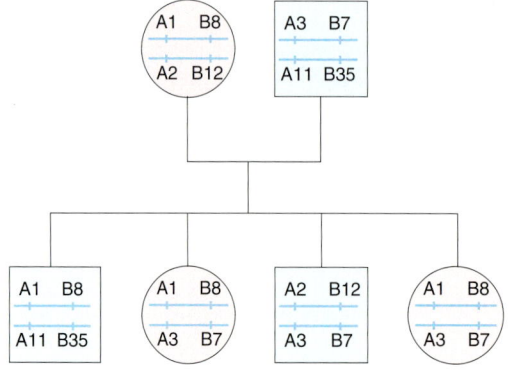

3 Family tree showing the HLA types

Immunological tolerance and autoimmunity

As early as 1900, the German microbiologist and Nobel Prize winner PAUL EHRLICH was fascinated by the fact that the body of an immunologically healthy human does not produce immunity against its own tissues. He called this concept the "Horror autotoxicus", which means the fear of self-destruction. EHRLICH realized, even then, that if this principle was disobeyed, severe problems would arise.

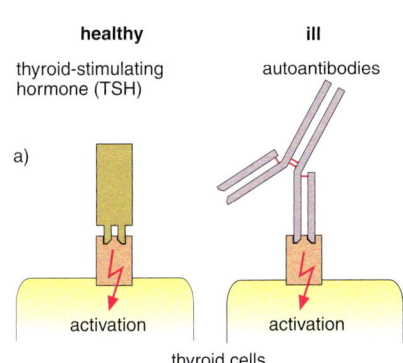

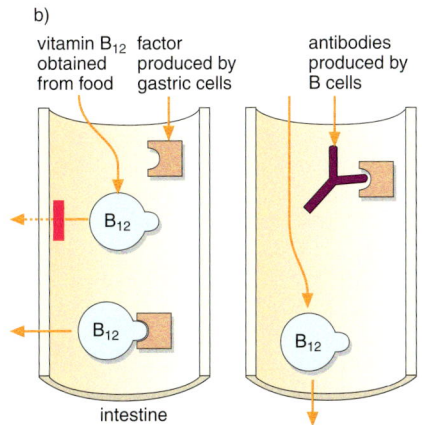

1 Thyroid hyperfunction (a) and pernicious anaemia (b)

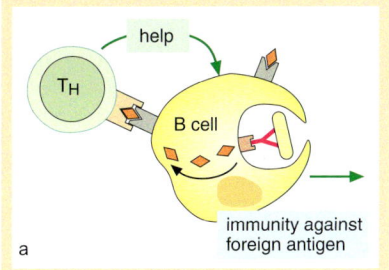

a

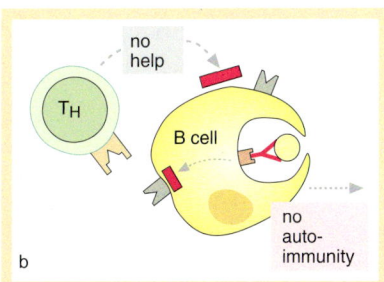

b

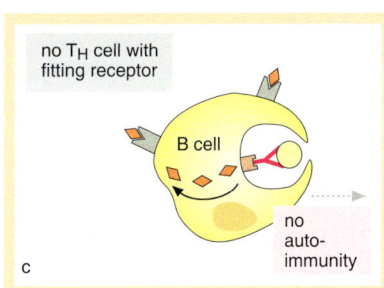

c

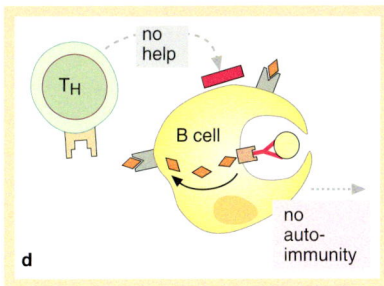

d

2 Immunological tolerance

Today, the molecular and cellular mechanisms of immunological tolerance and its demise are known. T and B cells and their antigen-receptor molecules are the key elements of the immune system (see page 86). If this activity is directed against the body's own tissue, we talk about *autoimmunity*. Antibodies directed against self-antigens are called autoantibodies. The specificity of the *autoantibodies* that are produced influence whether single organs (e. g. thyroid, pancreas) are destroyed or entire organ systems (muscle and joint system) are affected (e. g. inflammatory rheumatism).

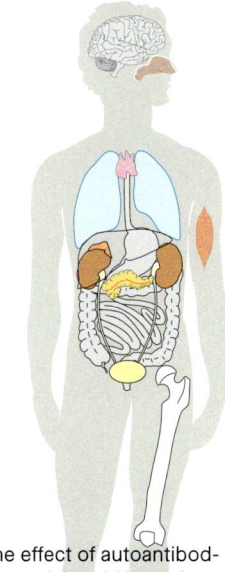

Tasks

① Describe the effect of autoantibodies in the case of thyroid hyperfunction and pernicious anaemia (fig. 1).

② The activation of B cells leads to the production of antibodies against foreign antigens (T cell-B cell cooperation, see page 86). Why are autoantibodies not produced against self-antigens in the situations shown in fig. 2b, c, d?

③ Helper T cells have to bind the presented antigen by using their T-cell receptor according to the lock-and-key principle in order to activate B cells to produce antibodies (fig. 3). Why do immunologically healthy individuals not develop autoimmunity? What mechanisms might be responsible for the development of autoimmune reactions in thyroid hyperfunction and pernicious anaemia?

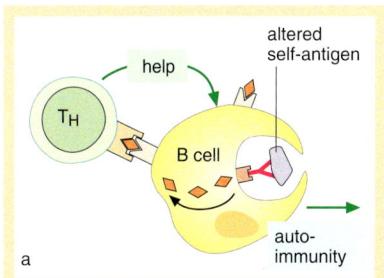

a

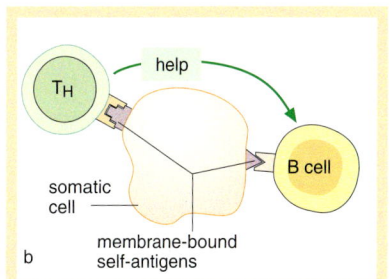

b

3 Destruction of immunological tolerance

AIDS

AIDS
Acquired
Immune
Deficiency
Syndrome

HIV
Human
Immunodeficiency
Virus

By 2002, the pathogen causing human immune deficiency syndrome (*HIV*) had killed about 20 million people. About 42 million people are infected worldwide and the number is rising daily by about 15,000. Every year, about 600,000 babies are infected during birth. In some developing countries, life expectancy has sunk to 36 years because of AIDS. In Germany, the number of registered newly infected individuals per year adds up to about 2,000. A shocking figure.

The first cases of AIDS were described in 1981: several young patients showed disease symptoms that were until then only known from elderly patients. They suffered from skin tumours and various infectious diseases. The immune system of these patients was shown to be significantly weakened. They died after a few months or years.

HIV: the AIDS pathogen

In 1983, the pathogen causing the immune deficiency syndrome AIDS was identified as the virus called HIV (*human immunodeficiency virus*). The HI viruses are spherical (Ø 0,1 μm). They have an outer envelope made of a lipid double layer that is derived from the outer membrane of the host cell. This envelope contains glycoproteins making the virus look like a pincushion (fig. 1). The HI virus belongs to the *retroviruses* and thus contains RNA as its genetic material. This viral RNA is located in a protein-containing capsid that also contains, amongst other molecules, the enzyme *reverse transcriptase*.

If HI viruses pass into the body fluids of a person, they bind according to the lock-and-key principle mainly to cells with the cell surface molecule CD4. These cells are helper T cells (but also monocytes, macrophages and specific nerve cells). The viral envelope fuses with the membrane of the host cell and the genetic information of the virus (RNA) reaches the cytoplasm. Then, reverse transcriptase converts the viral RNA to DNA (*reverse transcription*). This viral DNA integrates into the cellular DNA inside the host nucleus. The HI virus can thus remain in the organism for years as a pro-virus without being recognized by the immune system. However, if the infected helper cell becomes activated because of a different infection, the DNA of the pro-virus also becomes part of the activated transcription and translation. As a result the host cell produces HIV RNA and viral proteins. Finally, the helper T cell bursts and releases virus particles that enclose themselves during this process with the membrane of the host. The released HI viruses can now infect other immune cells.

Course of disease

During the course of an HIV infection, the number of helper T cells decreases dramatically. However, this cell type has a key position in the adaptive immune system (see page 86). Various stages are distinguished

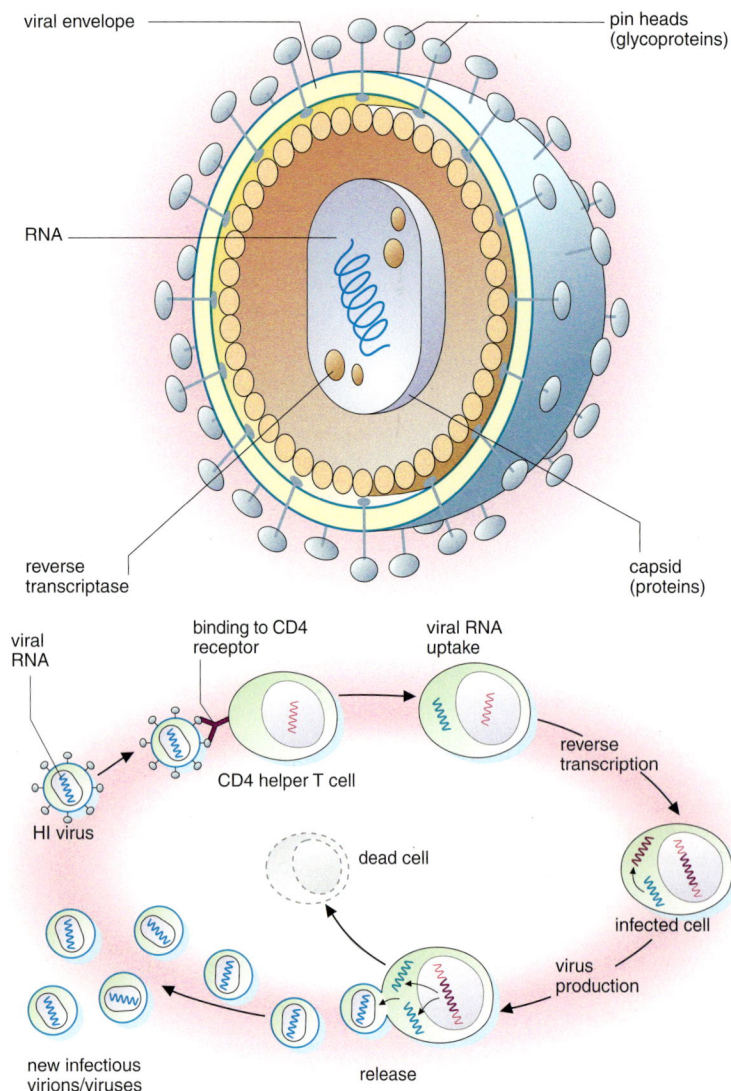

viral envelope — pin heads (glycoproteins)

RNA

reverse transcriptase — capsid (proteins)

viral RNA

binding to CD4 receptor — viral RNA uptake

HI virus

CD4 helper T cell

reverse transcription

dead cell

infected cell

virus production

new infectious virions/viruses

release

1 HI virus and its replication cycle

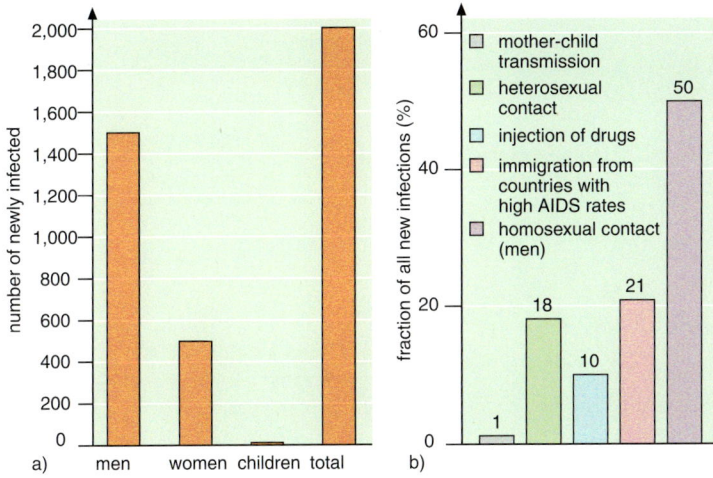

1 HIV: new infections (a) and transmission routes (b), Germany 2002

during an HIV infection. The early stage, within the first few weeks after infection, is represented by high fever and swollen lymph nodes. The body produces antibodies that can be detected and viral RNA (*HIV-positive*); however, the antibodies are not very effective. The symptoms fade and an HIV-positive person can live without symptoms for years, even though he can infect other people. Later, the number of helper T cells decreases and the immune system is increasingly weakened so that, in the third stage, even harmless infections cause diseases. The lymph nodes are continuously swollen. Only the final stage of a HIV infection is called *AIDS*. During this stage, infections (such as pneumonia, toxoplasmosis) and typical tumours (Kaposi's sarcoma, lymphoma) or changes in the brain lead to the death of the patient.

Therapeutic options

Today, an HIV infection cannot be cured. *Drugs* can slow down the course of the disease by inhibiting viral replication and stabilizing the number of helper T cells. This approach is focused on *reverse transcriptase*. It is inhibited by base analogues such as AZT (*azidothymine*). AZT resembles the base thymine and disturbs reverse transcriptase and hence the production of viruses. Therefore, the virus concentration in the blood sinks and the lymphocyte concentration rises. The drug has to be taken for the rest of the patient's life. In pregnant women, the infection of the unborn child can be prevented by AZT application.

The quest for an HIV vaccine has proved to be very difficult for several reasons: HI viruses are highly variable because reverse transcriptase makes many base pairing mistakes when converting RNA to DNA. Thus, new virus mutants are produced before the host organism can produce effective antibodies. Furthermore, the virus envelope causes trouble because it encloses the virus in a camouflaged sheath of host molecules that will not be attacked by the immune system. At the moment, antibodies are being sought that recognize the glycoprotein of the "pins"; this glycoprotein must be relatively stable because otherwise it could not bind to the receptor molecules on the helper T cells (fig. 94.1). In addition, researchers are trying to isolate antibodies from the blood of HIV-positive people who have been living for a comparatively long time without showing AIDS symptoms.

Risk of infection and protection against infection

As long as there is no effective treatment of AIDS available, the best protection is to avoid infection. Therefore, the transmission routes of the HI viruses must be determined: the virus appears in body fluids containing cells, such as lymph, breast milk, and semen or vaginal secretion. Blood contact carries the highest risk of infection. All blood destined for transfusions is tested for HIV today. It is possible to avoid direct contact with blood from, for example, accident victims by wearing gloves. Drug addicts can transmit the infection when sharing the same needle, which still contains traces of blood. In our society, new infections usually occur because of unprotected sex. The usage of condoms has significantly lowered the risk of infection. Only by acting responsibly can an individual protect himself/herself and avoid further transmission of the pathogen (fig. 1). Partners should only stop using condoms when both have been shown unequivocally not to be virus carriers. This is carried out with a HIV test that is performed as part of the routine, for example, during blood donation.

In everyday life, e. g. at work or in school, there is no risk of transmission. HIV-infected people should not be excluded and can take part in social life without being a danger to others. They need sympathy because of their difficult situation.

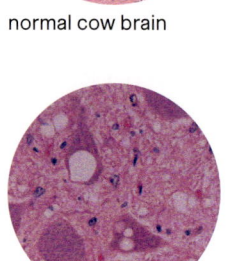

normal cow brain

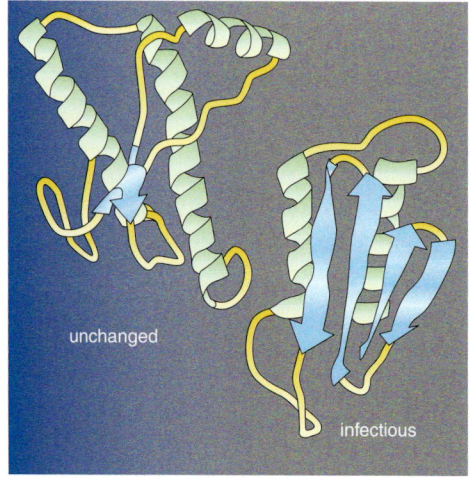

unchanged

infectious

1 Normal protein (left) and prion (right)

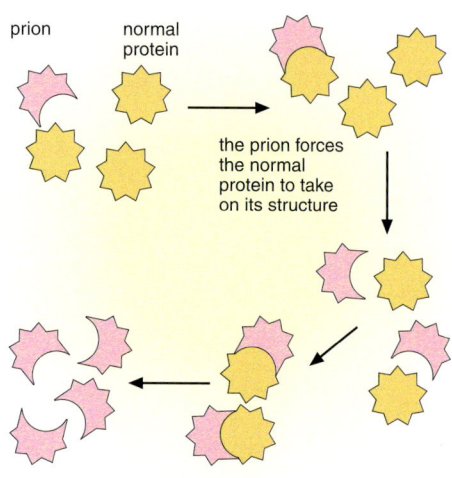

prion   normal protein

the prion forces the normal protein to take on its structure

2 Prions change the shape of normal proteins

spongiform BSE-damaged brain

bovis (Latin: cow)

spongiform, (Latin: like a sponge)

encephalopathy (Greek: brain disease)

Spongiform encephalopathy (spongy brain) in the animal kingdom

human: Creutzfeldt-Jakob disease (transmitted by cattle), kuru, Gerstmann-Sträussler-Scheinker syndrome

sheep: scrapie

domestic cat, chamois, kudu, antelope, oryx (no specific term for the disease)

Prion diseases: proteins become infectious

Prions belong to those discoveries that make scientists question whether they have really understood the basics of biology. Prions are proteins that can cause infectious diseases. In contrast to viruses, bacteria, fungi or parasites, they manage to multiply without nucleic acids. Until 1987, the year of the discovery of prions, no one had dreamt that proteins could be infectious.

Prions and BSE

BSE (mad cow disease) is an infectious disease in cattle that was first observed in 1984 in a cow in England. The animal was uncommonly aggressive and suffered from severe coordination problems, could not control its extremities, fell down all the time and died quickly. After more of these cases had occurred, the carcasses were examined systematically. In the brains, "holes" were found that were caused by massive cell death (see margin). Since the brain tissue resembled a sponge, the disease was called BSE (*bovine spongiform encephalopathy*). While searching for the agents causing BSE, a specific type of protein was found, namely *prions* (*proteinaceous infectious particle*). These prions were shown to be proteins made by the body itself, although they have a strongly changed spatial structure (see fig. 1). In experiments, this changed prion form was resistant to degradation by proteases. In 1987, STANLEY PRUSINER presented a theory of the molecular mechanisms involved in infectious prion diseases. According to him, infectious prions pene-

trate nerve cells and force normal proteins to change into their prion form (see fig. 2). The disease-causing prions multiply in the nerve cells with a domino effect. Their increasing concentrations disturb the cellular household and the nerve cells cease their function and die. Thereby, prions are released that infect further brain cells and kill them. The brain develops the structure of a sponge. PRUSINER'S theory of infectious proteins as a cause of BSE has been demonstrated to be true and he was awarded the Nobel Prize for medicine in 1997.

Several brain diseases in humans and animals are now counted amongst the *spongiform encephalopathies* that are caused by prions (see margin). It is especially dangerous for humans that the BSE pathogen can cross over species. Thus, we now believe that, in England, cattle were infected by eating food that contained the meat and bone meal of sheep that were infected by a spongiform encephalopathy (scrapie). Furthermore, some people seem to become infected by eating the material of cattle containing prions. This is the reason for beef being very strictly controlled today; indeed, brain cannot be used for the production of, for example, sausages any longer.

Theories of immunity

Pre-scientific theories

Ever since people have lived in close social organisations, they have repeatedly suffered from epidemics. Epidemics were described even in the old Egyptian dynasties. At that time, people believed (and in cultures lacking scientific knowledge, people still believe) that nature and humans are influenced by the magical powers of ghosts, demons and gods. Diseases were interpreted as the consequences of human sins or as a punishment for the breaking of taboos. In the Old Testament, God kills by using the plague, amongst others, and this disease was seen as a divine punishment. A person surviving the disease was said to have led a God-fearing life.

Task

① During the time of early Christianity, these perceptions changed. In addition to the idea of punishment, the idea of cleansing arose. Hereby, God uses a disease "to cleanse" people from the sins that they have committed. If a person had been freed from his sins, he would not be punished a second time by the disease. Explain why this idea approaches our present day understanding of acquired immunity after an infection has been overcome or after vaccination.

Historical reports

"I say that every person dries out during the time from his birth to old age. Because of this, the blood of children is wetter than the blood of young men and, of course, is wetter than the blood of old men. ... Pox occurs when the blood becomes rotten and ferments so that excess fluid is secreted and the blood of children, which is like grape-juice, turns into the completely ripe blood of young men, which is like wine. The blood of old men can be compared to wine that has lost its strength and starts to be stale and bitter. Pox can be seen as fermentation accompanied by the fizzing sound that is present in grape-juice. And this is the reason why children, especially male children, seldom escape this disease because it is impossible to stop the change of the blood from one state to the other as it is impossible to stop grape-juice from turning into wine."
(Islamic physician Rhazes, 10th century).

"Then said Yahweh to Moses: go to the pharaoh and tell him: so says Yahweh the god of the Hebrews: release my people so they can serve me. If you refuse to let them go, then the hand of Yahweh will come over your animals on the fields, and over the horses, the donkeys, the camels, the cattle and sheep like a horrible plague. But Yahweh will make a difference between the animals of the Israelites and those of the Egyptians. No animal belonging to the Israelites will die ... "
It happened as prophesied; the pharaoh remained stubborn, and the animals of the Egyptians died but not the animals of the Israelites.
(Bible, Book of Exodus: liberation from Egypt, 3. The Egyptian Plagues, 5th plague: disease of livestock)

Tasks

② People rely on their ability to make assumptions about the events that occur around them, finding explanations and making predictions. What reasons are given for the diseases that are mentioned in both texts?

③ Were the people able to test explanations of the reasons for the disease? Give explanations for your opinion.

Scientific theories

Between 1930 and 1950, chemists successfully analysed the structure of antibodies; they characterised antigens and those forces enabling the reversible binding of the antigen and antibody according to *lock-and-key principle* (electrostatic attraction, hydrogen bonds, hydrophobic bonds, van der Waals forces). Small molecules that do not occur in nature (*haptens*) were synthesized in test tubes and linked to large *carrier* proteins. When injected into mammals, the carrier-hapten-complex provoked antibody production. The antibodies were isolated and they were shown to recognise haptens specifically.

Stencil theory

In 1940, the biochemist Linus Pauling proposed the theory that antigens function as a stencil for the production of exactly fitting antibodies. These antibodies should literally form around the antigens during their synthesis and be a negative imprint that persists after the release of the antibodies into the body fluids.

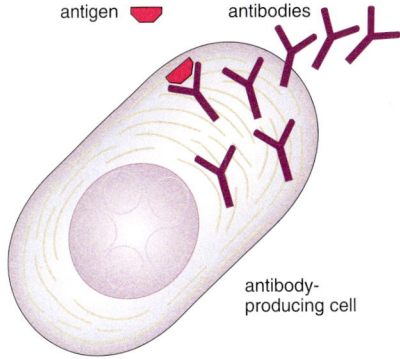

antigen   antibodies

antibody-producing cell

Task

④ Scientific theories differ from pre-scientific theories because they are disputable. Disprove the stencil theory of antibody production based on today's state of knowledge regarding the genetics of antibody diversity (see page 89).

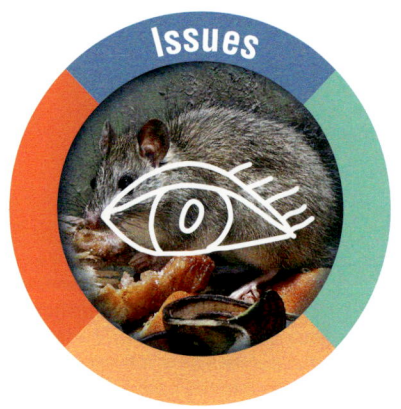

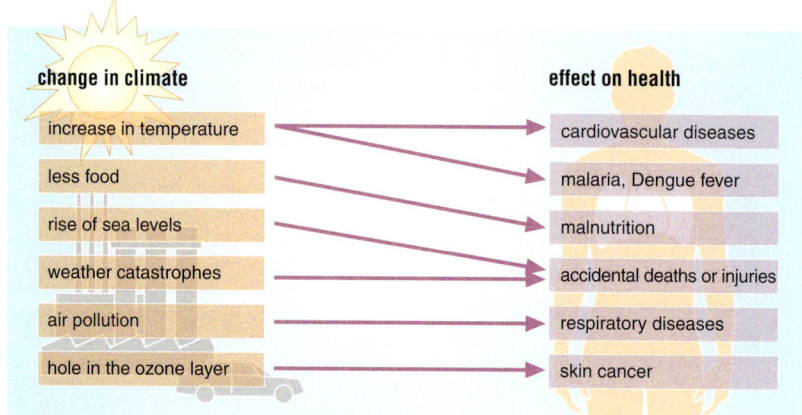

change in climate	effect on health
increase in temperature	cardiovascular diseases
less food	malaria, Dengue fever
rise of sea levels	malnutrition
weather catastrophes	accidental deaths or injuries
air pollution	respiratory diseases
hole in the ozone layer	skin cancer

Global health

Humans have always been exposed to the influence of diseases, injuries and accidents. Especially infectious diseases such as *malaria*, *plague* or *cholera* have had devastating effects. The catastrophic plague endemic in 1348 resulted in the depopulation of some areas. Today, infectious diseases are number one on the list of causes of death, causing globally 17 million deaths per year, followed by 15 million deaths attributable to cardiovascular diseases and 7 million deaths resulting from cancer. According to the *World Health Organization* (WHO) in Geneva, AIDS, tuberculosis and malaria cause most deaths amongst the infectious diseases. The reason for this is the present lack of any effective vaccines against these diseases. Furthermore, new diseases develop continuously, such as BSE, SARS or bird flu. Many people are also victims of famines: e. g. China (1958 to 1960) 30 million deaths. In 2020, it is expected that diseases resulting from tobacco consumption will exceed every other disease.

World Health Organization

 The WHO is a special organisation within the United Nations.

Extract from the constitution of 1946:

— Health is a state of complete physical, mental and social well-being and not merely the absence of disease or physical handicaps.
— The enjoyment of the highest attainable standard of health is one of the fundamental rights of every human being without distinction of race, religion, political belief, economic or social condition.
— The health of all peoples is fundamental to the attainment of peace and security and is dependent upon the fullest co-operation of individuals and states.
— The achievement of any state in the promotion and protection of health is of value to all.
— Unequal development in different countries in the promotion of health and control of disease, especially infectious disease, is a common danger.
— Governments have a responsibility for the health of their peoples, which can be fulfilled only by the provision of adequate health and social measures.

Take a stand on the following statement: There are no healthy people but only badly examined patients

How much can the WHO influence the politics of its member states with regard to health?

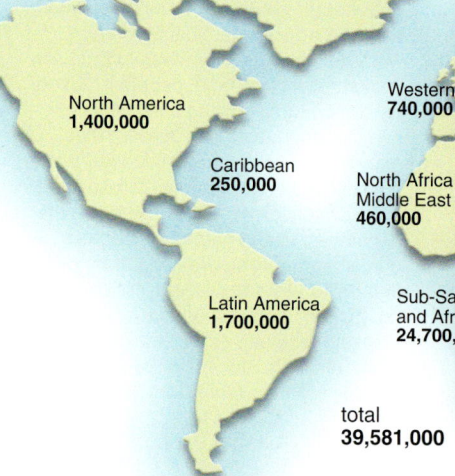

Number of HIV infected (2002); source: UNAIDS

North America
1,400,000

Caribbean
250,000

Latin America
1,700,000

Western E
740,000

North Africa a
Middle East
460,000

Sub-Sah
and Afric
24,700,0

total
39,581,000

Compare the distribution of AIDS, malaria and tuberculosis. State biological and social reasons contributing to their respective distributions.

Environmental politics and health

Health Impact Assessment (HIA) = (test for health tolerance)
"HIA is a practical approach used to judge the potential health effects of a policy, program or project on a population, particularly on vulnerable or disadvantaged groups. Recommendations are produced for decision makers and stakeholders, with the aim of maximizing the proposals for positive health effects and minimizing the negative health effects."

Environmental pollution directly and indirectly influences global health. Develop a concept for applying HIA to political actions.

Health prevention

The English physician EDWARD JENNER (1749 – 1823) observed that people were protected from smallpox if they had previously suffered from *cow pox*. As evidence, he infected a child with cow pox and 6 weeks later with smallpox — the child did not become ill. This is where *vaccination* obtained its name (Latin: *vacca* means cow). Smallpox vaccination has been globally successful and the last smallpox case was registered in 1977.

Obtain information about the vaccinations recommended in Germany and check your personal protection as provided by vaccination. Which diseases are notifiable (must be reported to authorities) in Germany?

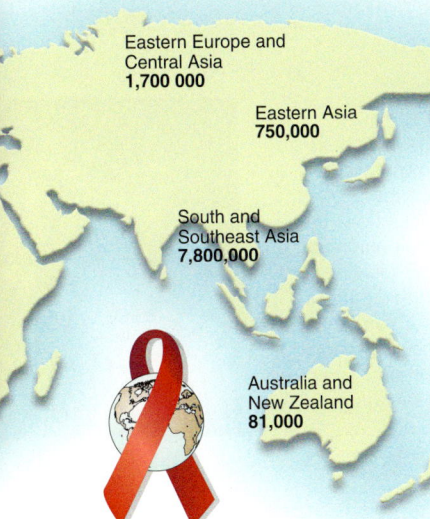

Eastern Europe and
Central Asia
1,700 000

Eastern Asia
750,000

South and
Southeast Asia
7,800,000

Australia and
New Zealand
81,000

Effects of globalisation

At the beginning of 2003, an atypical pneumonia featuring high fever, breathlessness, a heavy cough and sore throat dramatically spread in China. Before that time, such SARS cases (*severe acute respiratory syndrome*) had been reported only occasionally in Vietnam, the Philippines, Indonesia and Canada. The cause of SARS is a virus that until 2003 was harmless for humans. Transmission occurs mainly in the respiratory tract (*droplet infection*). SARS viruses can spread from animals to humans if they live together very closely. The pathogen was probably carried by a physician from South China to Hong Kong from where the disease spread almost all over the world. By July 2003, 8,445 people were infected worldwide and 812 had died.

Apart from medical effects, the epidemic also caused severe economic effects, especially concerning holiday and flight traffic.

Compare a current disease such as SARS with a medieval one such as the plague with regard to distribution, diagnosis, therapy and prophylaxis.

Diseases such as BSE or SARS also have economic consequences for the affected countries. Explain.

What is the importance of national and international cooperation when fighting SARS?

Travel advice for tropical countries
— Only eat peeled or boiled, fried or barbequed food (cook it, peel it or forget it!)
— Drink only bottled drinks (no tap water not even for brushing teeth, no ice cubes and no ice cream).
— Wear long-sleeved clothes in the evenings and long trousers against the danger of insect bites.
— Avoid swimming or bathing in rivers or lakes in Africa and South America (*danger of bilharzia*).
— Tuck the ends of mosquito nets well underneath the mattress.
— When staying in the tropics longer than 3 months, have your blood, urine and faeces tested, even if you have no symptoms.
— ...

Develop a travel plan and travel advice for a trip to the tropics, e.g. Gambia or Brazil, by taking health aspects into account.

Mark, on a world map, all those countries that have been visited by your fellow pupils. Compare the result with the travel experiences of your (grandparents and) parents when they were of the same age as you are. State the effect of tourism on the spread of infectious diseases.

It is assumed that even explorers such as Columbus and Cook contributed to the spread of diseases. What can you find out about this?

Poverty and health

"East and Southeast Asia have reduced their division of malnutrition from 52% in 1970 to 32% in 1992, whereas it has increased in the same time from 26% to 30% in South Asia, and in Africa south of the Sahara it jumped from 11% to 26%.

During the same time, the economic growth rate in East and Southeast Asia were 4% to 8%, whereas in South Asia, the growth rate hardly reached 3% and, in Africa south of the Sahara, the gross domestic product decreased ...

One fifth of all people have no access to modern health services and half of the world's population cannot get regular access to basic drugs. ... We have known for a long time that poverty causes bad health and that bad health prolongs poverty. ..."
DR. GRO HARLEM BRUNDTLAND, 1999,
Chief executive of the WHO

Many diseases such as malaria, AIDS, tuberculosis or sleeping sickness can be eased with drugs or cured. Why do developing countries profit only very little from this?

Reproduction

In publishing, reproduction is the act of copying a document or image by printing techniques. In the arts, a reproduction could be called a copy of an original picture. In acoustics, reproduction is the high fidelity production of sound. In cabinet-making, a reproduction is a copy of an antique piece of furniture. In all cases, something is "re-created" or "re-produced" and, for such reproduction, an external force is required.

In biology, the term "reproduction" is used as synonym for the process of generating offspring. Organisms, in contrast to inanimate nature, have the ability to self-duplicate: life creates life. The result, which is determined by the lifespan of an individual, is a sequence of generations that are capable of being modifiied and that can thus evolve. In this process, several mechanisms take part that can be influenced by humans by using selective breeding, gene engineering and reproductive medicine.

Recombination and diversity

During meiosis, the genetic information of the parent is newly combined by the random distribution of the homologous chromosomes. Within a chromosome, genes can be rearranged after crossing-over. Bacteria achieve similar results by using so-called parasexual processes during transformation, transduction and conjugation. In all cases, the result is an increase in the number of genotypes that lead, by selection, to a better adaptation of the organisms to their environment.

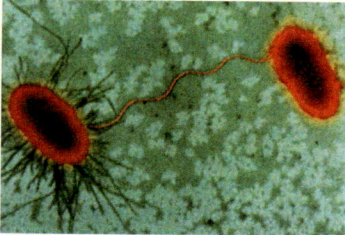

Recombination and diversity

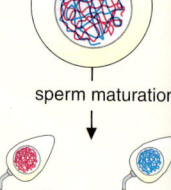
sperm maturation

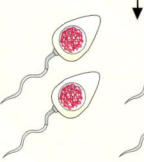

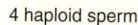
4 haploid sperm

Replication and mutation

The "identical" replication of DNA is not always a perfect copying process: small mistakes by the polymerase that pairs complementary nucleotides to the original strand can lead to point mutations. Other mistakes can cause frame shifts, the abrupt ending of replication or nonsense sequences. Even though genetic information is mainly preserved, variants can thus be created that represent the "playthings" of evolution.

Replication and mutation

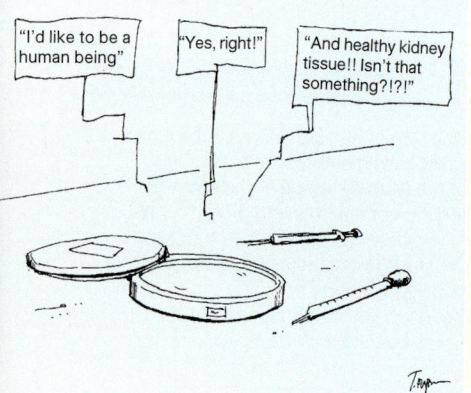

In the reproduction lab

"I'd like to be a human being"

"Yes, right!"

"And healthy kidney tissue!! Isn't that something?!?!"

Asexual reproduction

Unicellular organisms divide repeatedly; plants form runners, bulbils or axillary buds. Such *asexual reproduction* is uncommon in higher animals, although artificially genetically identical organisms can be created by cloning in mammals. Asexual reproduction in which no recombination of the genetic material is possible should not be confused with unisexual reproduction (*parthenogenesis*), which is used by aphids in order to carry out mass reproduction.

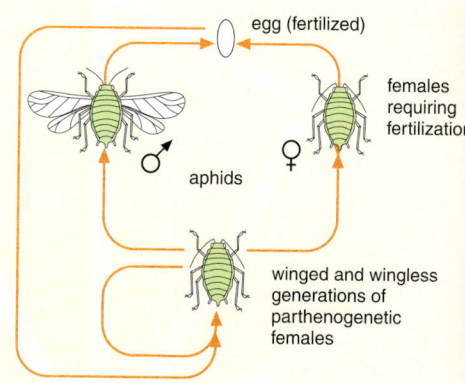
egg (fertilized)

females requiring fertilization

aphids

winged and wingless generations of parthenogenetic females

Asexual reproduction

● **Think about connections**   ● **See connections**   ● **Develop connections**

Sexuality

Reproductive strategies

Parents „invest" in their offspring at great expense. Specific behavioural strategies ensure the most effective use of limited resources. This can, for example, lead to a species reproducing only seldomly during its lifetime but investing extensively in brood care. Other species that expend less can have more offspring more often. Selection favours the best possible compromise. Ecological factors are almost always crucial in this respect.

Agaves live in hot dry habitats and reproduce asexually. Flower or fruit production occurs only during extraordinary wet years. This enables the young flowers to germinate successfully. The seeds are produced in excess so that they are not wiped out by being eaten.

Sexuality

In sexual reproduction, a large stationary female gamete is fertilized by a small motile male gamete. The number of viable reproductive units (e. g. sperm or fertile pollen, egg cells or germinable ovules) determines the *reproductive efficiency* of an organism. Selection in contrast determines the *reproductive success* and describes the number of surviving offspring. This in turn changes the allele frequency of the various genes *(reproductive fitness)*. Offspring who help with the upbringing of their siblings also have a reproductive fitness.

Reproductive strategies

Tasks

1. The so called big bang strategy is employed by some bamboo species: they live for many years in vegetative growth phases and then invest all their energy in one single reproductive phase, dying thereafter. State advantages and disadvantages.
2. The males and females of the Pacific salmon feed on their body resources while travelling to their spawning grounds and die after spawning. Describe the costs and benefits of this strategy.
3. Assess the reproductive fitness these behaviours: territory defence, brood care or the creation of specific nests to raise offspring.
4. Reproduction as in "production of offspring" does not necessarily mean duplication. Explain this.
5. The various forms of asexual (vegetative) reproduction ensure that an individual passes on copies of all its genes. Compare the advantages and disadvantages of this with sexual reproduction.
6. Fitness as described by Darwin is not measured as the number of offspring produced but as the number of surviving offspring that can themselves produce offspring. Different reproductive strategies have emerged during evolution. Explain the connections.
7. Florida Scrub Jays have so called "helpers at the nest" that do not reproduce. Under what circumstances can this unselfish (altruistic) behaviour have a beneficial effect for reproductive fitness?

Sex determination

Usually, sex is irreversibly determined when an egg cell is fertilized *(genotypic sex determination)*. In rare cases, external factors such as temperature can be the determining factor. In spoon worms, contact of the larvae with a female determines its development as a male. In coral gobies, ecological factors are crucial: on the Australian coast, this fish rarely finds suitable corals that allow breeding in pairs. The young fish therefore remain sex neutral. Breeding places that become available are occupied immediately by a fish and its partner becomes the opposite sex.

Sex determination

Think about connections ● **See connections** ● **Develop connections**

Information and communication

Organisms receive, save and process information and communicate with each other. Precondition for this are a common language and suitable receiver, storage and sending mechanisms.

The word "information" has various uses in everyday language: objects such as a CD *contain* information, and organisms *receive* information. In biology, we define information as a message that is made up of a spatial or chronological sequence of signals that causes a certain effect. Communication is a mutual transfer of information that is adapted to the communicating partners. It can occur between organisms but also within an organism and inside a cell.

Sender and receiver

In a communication chain, the sender transmits an encoded signal that not only must arrive at, but must also be decoded by the receiver. This means that signals and processing mechanisms must be compatible. One example is *pheromones*. These substances serve in the communication between individuals of one species. For example, *bombicol* — the sexual attractant of the silkworm moth — is only produced by the females. The appropriate receptors occur in the large antennae of the male moth and react even to minute concentrations of this substance.

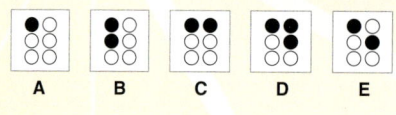

A   B   C   D   E

Encoding/Decoding

information → encoding → signals → decoding → information

sender (also transmitter)   receiver

code

A   B   C   D   E

Sender and receiver

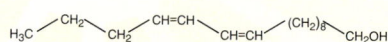

structural formula of bombicol

pheromone receptors on a sensory hair

Encoding/Decoding

A code involves specific instructions for converting information from one language to another. Examples are the Morse code, the encoding of texts in bits and bytes, and the telephone. With the genetic code, information in "the language of nucleic acids" is translated into "the language of amino acids".

By encoding information, transfer can be made safer. Nerve cells send messages encoded by frequency and process their modulation by amplitude.

Redundancy

The term "redundancy" is derived from information theory (Latin: *redundantia* = surplus) and describes the informative surplus of a message. In communication engineering, high redundancy protects a message from disturbances during transmission.

In genetics and evolution, redundancy is the multiple existence of similar signal structures. This ensures stability and reliability. Parsimoniously (according to economical principles), a minimal number of participating factors is used. Only five elements make up DNA, four bases encode the entire variety of all species, and about 20 amino acids form all proteins.

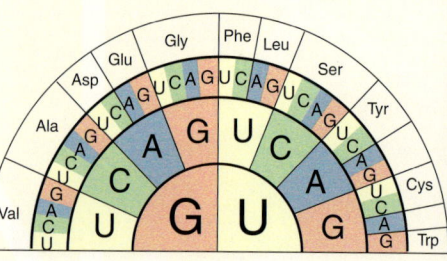

Redundancy

● **Think about connections**   ● **See connections**   ● **Develop connections**

Flow of information within a cell

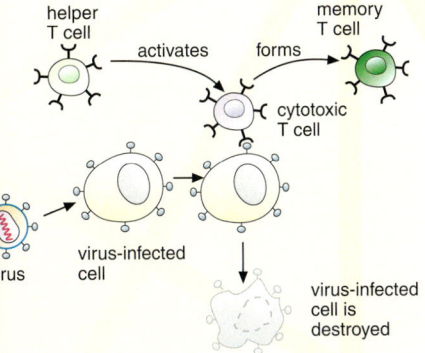

Flow of information within a cell

Within a cell, continuous information transfer occurs from organelle to organelle via various chemical substances. Adjacent cells within a tissue can also communicate with each other. Special cell-cell contacts make the exchange of small molecules possible. In the immune system, various cells communicate with each other either by signalling substances or by direct cell contact.

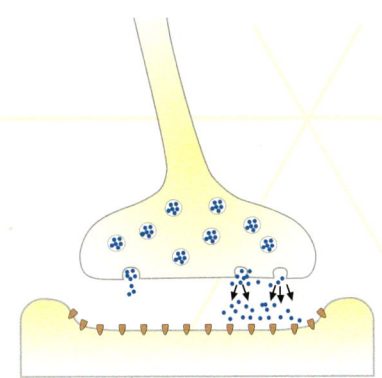

Nervous and hormone systems

The nervous system is used for fast and precise communication. Not only electronic signals, but also chemical signals are employed. The interaction of afferent and efferent routes influences the flow of information by means of feedback mechanisms. In addition to the fast effective transport of information by the nervous system, hormones are used in the blood stream for long-term communication. They have an effect on the corresponding receptors of their target organs.

Nervous and hormone systems

Communication

The term "language" applies primarily only to verbal communication between people. In the broader sense, it also includes symbolic languages (sign language) or the language of animals (e.g. the dance of the honey bee or body language). In social organisations, optical and acoustic signals for communication are important for social cohesion or for collective activities. Plants also communicate with each other or with animals, for example, by releasing warning substances when being attacked by pests.

Communi-cation

Tasks

1. Think about one example each from behavioural biology, neurobiology, metabolic biology and genetics showing how the sender (transmitter) and receiver are adapted to each other.
2. Short waves and medium waves in broadcasting have a significantly worse quality than very high frequency stations. Find an explanation.
3. Present different encoding mechanisms for information transfer in the nervous system.
4. Plants save information regarding light exposure conditions in the form of phytochrome molecules. Give details of the function of this molecule and explain how information is processed by it.
5. Books, movies, a CD or DVD are examples for technical storage media. Compare these data carriers to biological information carriers regarding the kind of information, its storage and its replay.
6. Justify the statement that nucleic acids are used for both information transfer and information storage.
7. State similarities and differences between the nervous and hormone systems. Compare the form of information transfer with cellular communication or with communication between individuals.
8. Why is it incorrect to refer to the sounds of "talking" parrots as a language?
9. Discuss the function of signalling colours during the courtship of many animals. Compare this with the function of warning colours.

Think about connections   ● **See connections**   ● **Develop connections**

Ethical assessment: basic tools

When assessing topics such as "cloning — reproductive medicine", social, legal and ethical questions have to be taken into account, in addition to pure expert knowledge. Thus, scientific knowledge (*descriptive dimension*) can be linked with the assessment of human action (*normative dimension*). This should lead to answers to the question regarding whether it is morally good or bad to act as intended. The ability to make a reasonable decision between two values, even in a dilemma, requires the ability to judge morally. Some "basic tools" can help in such situations.

Syllogism

1. "All techniques that serve to provide a childless couple with a child are good."
2. "Reproductive medicine can fulfil the desire of childless couples to have children."
3. "Reproductive medicine fulfils the desire of childless couples. It is therefore morally good and should be used."

The three steps represent the basic pattern of ethical reasoning:
1. *normative premise*,
2. *descriptive premise*,
3. *conclusion*.

The normative premise represents a value that is considered essential by the judging person. The value shows a direction of the purpose (having children is seen as a high value); the term "norm" is seen as the accepted standard.

Deontological reasoning

Deontological ethics assume a direct moral duty. The validity of statements such as "You should not lie!" is inherent. Some actions are wrong no matter what consequences follow from them. An understanding of morals and the duty to act according to them can be tested: an individual maxim (principle or rule) is tested by asking whether it is desirable that all other people act accordingly (*categorical imperative*). If a maxim fails this test, a strict ban to act in this way is the result.

Utilitarian/consequential reasoning

Utilitarian/consequential ethics are based on the idea that individual happiness is transferable to other people and that happiness is the criterion for all actions. Which social state is the better one can be determined when assessing the sum of all individual happiness. An action is considered good if its consequences are good for human happiness and it can maximise the "sum of happiness" among all persons. An individual action is considered good if it causes the best common social state when compared with all other possible alternatives.

Reasoning strategies

Two reasoning strategies can be distinguished: those that judge the consequences and those that focus on absolute (categorical) rules (duties, instructions and bans). When assessing the consequences of cloning and reproductive medicine, it is very important to consider "social justice", "cultural climate", "economics", "democracy", "inner nature" and "the health of the human body and soul" and, at the same time, to protect "the integrity of a human being or a person" (human dignity).

Ethical analysis

An ethical analysis can be divided into the following steps (compare Hössle, C., 2001):

1. Identify and formulate the decision or the dilemma.
2. Identify the actions that could be taken in this situation.
3. Analyse the actions and list them according to the different values that are addressed.
4. Make a decision based on reasons for one action while taking its consequences into consideration.
5. Analyse the reasons for your decision and relate them to the ethical theories.
6. Describe the consequences of an individual decision.

References

Hössle, C.: Moralische Urteilsfähigkeit. Studien-Verlag. Innsbruck 2001
O'Neill, O.: Autonomy and Trust in Bioethics. Cambridge University Press, Cambridge, 2002
Warnock, M.: An intelligent person's guide to ethics. Duckworth, London, 2006

Production of insulin by genetic engineering

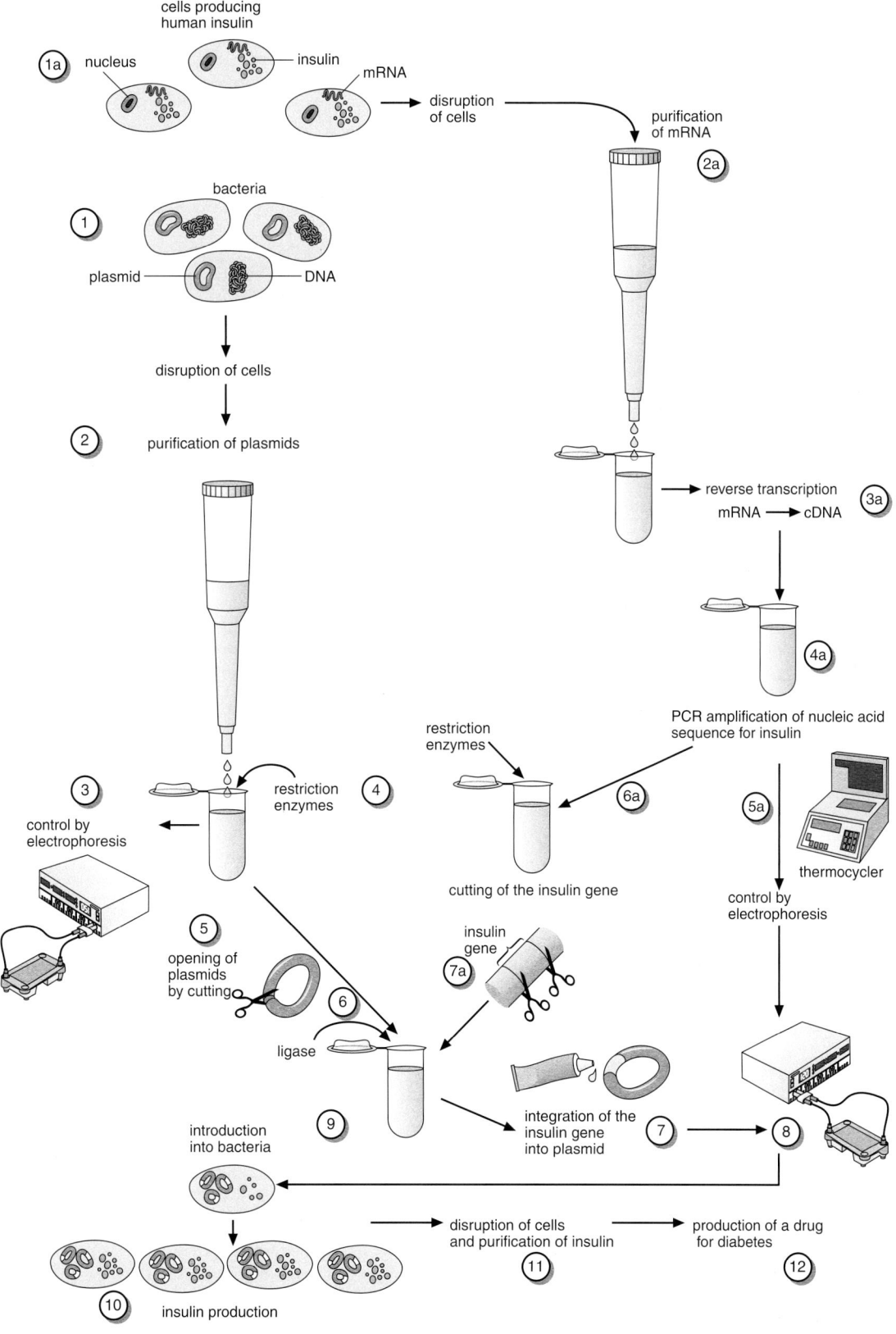

cells producing
human insulin

1a — nucleus — insulin — mRNA — disruption of cells — purification of mRNA — 2a

bacteria

1 — plasmid — DNA

disruption of cells

2 — purification of plasmids

reverse transcription
mRNA → cDNA — 3a

4a

PCR amplification of nucleic acid
sequence for insulin

restriction enzymes

3 — control by electrophoresis — restriction enzymes — 4 — 6a — cutting of the insulin gene — 5a — thermocycler — control by electrophoresis

5 — opening of plasmids by cutting

insulin gene — 7a

ligase — 6

9 — integration of the insulin gene into plasmid — 7 — 8

introduction into bacteria

10 — insulin production — disruption of cells and purification of insulin — 11 — production of a drug for diabetes — 12

Actin filaments [ˌæktɪn ˈfɪləmənts]
Long thin proteins made up of globular subunits. They are a part of the cytoskeleton.

Agglutination [əˌgluːtɪˈneɪʃn]
The clumping of blood cells or bacteria brought about by bridging molecules (e.g. antibodies).

AIDS (**a**cquired **i**mmune **d**eficiency **s**yndrome / acquired immunodeficiency syndrome
[eɪdz] [əˌkwaɪəd ɪˌmjuːn dɪˈfɪʃənsɪ ˌsɪndrəʊm/əˌkwaɪəd ɪˌmjuːnəʊdɪˈfɪʃənsɪ ˌsɪndrəʊm]
A disease of the immune system caused by the human immunodeficiency virus (HIV), which has caused 20 million deaths up until 2002 worldwide. Infected people are able to transmit the disease, even before symptoms of the illness, appear; the infection spreads via direct contact with body fluids.

Albino [ælˈbiːnəʊ]
An individual with white skin/fur and red eyes caused by a hereditary deficiency of skin and hair pigment. Albinos occur within birds and mammals including humans.

Allele [əˈliːl]
One of the many versions of a *gene*. *Diploid cells* of the body contain two alleles, and *haploid gametes* one.

Allergy [ˈælədʒɪ]
An abnormally high sensitivity of the immune system to certain substances called allergens.

Alternation of generations
[ˌɒltəˌneɪʃn əv ˌdʒenəˈreɪʃnz]
Life cycle of a species in which alternating generations use different kinds of reproduction; this could be sexual and asexual reproduction or *parthenogenesis*.

Amino acid [əˌmiːnəʊ ˈæsɪd]
Organic acid containing a carboxyl group (-COOH) and an amino group (-NH$_2$). Basic building block of proteins ("essential" amino acids cannot be synthesized by the organism itself).

Antibiotic [ˌæntɪbaɪˈɒtɪk]
A substance that is produced by microorganisms and that inhibits the growth of other microorganisms, e.g. penicillin, chloramphenicol.

Antibodies [ˈæntɪˌbɒdɪz]
Molecules that occur in the body fluids of vertebrates and that can bind specifically to other molecules *(antigens)*.

Anticodon [ˈæntɪˌkəʊdɒn]
A special sequence of three bases on a tRNA molecule; these bases pair with the complementary three-nucleotide codon of an mRNA molecule during protein synthesis.

Antigen [ˈæntɪdʒən]
A substance that stimulates the immune system to produce *antibodies*.

ATP (adenosine triphosphate)
[ˌeɪtiːˈpiː] [əˌdenəziːn traɪˈfɒsfeɪt]
The most important energy storage and energy transport molecule. It is made from ADP and phosphate by using energy. Its breakdown releases 30.5 kJ / mol and drives endergonic reactions. It can also transfer phosphate to other substances converting them to high-energy substances and making them more reactive.

Autoimmune disease
[ˌɔːtəʊˈmjuːn dɪˌziːz]
A disease in which an immune reaction occurs against the body's own cells. Diabetes and rheumatic diseases are autoimmune diseases.

Autoradiography [ˌɔːtəʊˌreɪdɪˈɒgrəfi]
A technique that uses X-ray film to visualise radioactively labelled molecules in tissue sections or cell spreads. A sensitive method for the localisation of substances or for the detection of the synthesis, transport or turn-over of substances.

Autosome [ˈɔːtəʊsəʊm]
A chromosome, which in contrast to *gonosomes*, does not carry genes required for sex determination. *Diploid cells* of somatic cells (cells of the body) have two homologous copies of each autosome.

Bacteriophage (phage)
[bækˈtɪərɪəfeɪdʒ] [feɪdʒ]
A virus that infects bacteria. It uses the machinery of the bacterium for its own reproduction. Phages contain *proteins, DNA* or *RNA*.

Benign [bəˈnaɪn]
Not cancerous. Benign tumours do not spread to tissues around them or to other parts of the body.

Biomembrane [ˌbaɪəʊˈmembreɪn]
Membrane bilayer enclosing cells or compartmentalizing cells into small reaction spaces. It is composed of two lipid layers and proteins.

Biotechnology [ˌbaɪəʊtekˈnɒlədʒɪ]
The use of biological processes or living organisms in technological applications and industrial processes.

Blastula [ˈblæstjʊlə]
Early stage of the development of a multicellular animal in which a single layer of identical-appearing cells surrounds a fluid-filled cavity, forming a hollow ball.

Blotting [ˈblɒtɪŋ]
A common laboratory procedure in molecular biology in which molecules (*proteins, nucleic acids*) after gel electrophoresis are transferred directly and therefore in the same arrangement from the gel to a membrane for further scientific analysis.

Breakage and reunion
[ˌbreɪkɪdʒ ən ˌriːˈjuːnɪən]
The breakage of a group of originally linked genes, i. e. genes that are located on the same *chromosome*, and their subsequent rejoining. Breakage and reunion can occur during *crossing over* at *meiosis* or because of a chromosome mutation.

Cancer [ˈkænsə]
A term that collectively describes malignant tumours in the tissues of the body. Cancerous tumours can spread throughout the body (forming *metastases*) in contrast to benign tumours.

Cell [sel]
The basic structural and functional unit of all organisms. Cells only arise from other cells. Two main types of cells are distinguished: the *prokaryotic cell* and the *eukaryotic cell*. Organisms consisting of eukaryotic cells can be uni- or multicellular.

Cell cycle [ˈsel ˌsaɪkl]
The series of steps through which a cell passes in order to duplicate its genetic material (DNA synthesis) and split into two new cells. This process is strongly controlled and regulated.

Centriole [ˈsentrɪəʊl]
One of a pair of small cell organelles in the *centrosome* region of animal.

Centrosome ['sentrəsəʊm]
A small region in the *cytoplasm* of *eukaryotic cells* important for cell division. The centrosome of most animal cells contains a pair of *centrioles*.

Chloroplast ['klɒrəplɑːst]
A cell organelle containing chlorophyll. It is involved in photosynthesis (see *plastid*).

Chromosome ['krəʊməsəʊm]
The main structures carrying the hereditary information of a cell. Chromosomes contain genes *(DNA)* in a specific arrangement and are duplicated identically during cell division *(replication)*. In *eukaryotes*, they become visible in the light microscope during cell division. In the broader sense, the circular DNA of a prokaryote can be referred to also as a (bacterial-) chromosome.

Chromosome mutation
[,krəʊməsəʊm mjuː'teɪʃn]
A change in the hereditary information caused by breakage or fusion. A change in the gross structure of a chromosome. Parts can be lost (deletion), moved to another location (translocation), duplicated (duplication) or turned by 180° and reinserted (inversion).

Cleavage ['kliːvɪdʒ]
First phase of the embryonic development of a multicellular organism during which the *zygote* becomes a multicellular ball *(blastula)* after repeated cell divisions. Depending on the amount of yolk in the egg, the cleavage can be total in egg cells with little yolk or partial in egg cells rich in yolk.

Cloning ['kləʊnɪŋ]
The process of making duplicates that are genetically identical copies of the original cell; also applies to the replication of single genes (see PCR).

Codominant [,kəʊ'dɒmɪnənt]
Two different *alleles* of the same gene that are both fully expressed in the phenotype of a diploid organism; if one of them occurs together with a recessive allele, it dominates the phenotype (see *inheritance*).

Codon ['kəʊdɒn]
A sequence of 3 nucleotides on an mRNA segment (base triplet) that codes for an *amino acid*.

Coenzyme ['kəʊ,enzaɪm]
An organic cofactor molecule that supports enzymes by temporarily binding to electrons, protons or parts of substrates during enzymatic reactions. (Coenzymes are also sometimes referred to as cosubstrates).

Complement system
['kɒmplɪmənt ,sɪstəm]
Part of the immune system. A group of about 20 serum proteins that react nonspecifically with foreign cells either to destroy them or to prepare them for degradation by macrophages.

Conformation [,kɒnfɔː'meɪʃn]
The spatial arrangement of a chemical compound.

Conjugation [,kɒndʒʊ'geɪʃn]
The process by which protozoa or bacteria in close proximity exchange genetic information.

Copulation / Mating
[,kɒpjʊ'leɪʃn] ['meɪtɪŋ]
Sexual union of the sex organs of a male and a female individual for the direct transfer of sperm cells into the female reproductive organs. This ensures insemination and subsequent internal fertilisation of the egg cell.

Crossing over [,krɒsɪŋ 'əʊvə]
The reciprocal breaking and rejoining of homologous chromosomes during meiosis resulting in the exchange of chromosomal segments. Crossing over leads to the recombination of genetic information by a reciprocal exchange of parts of the chromosome.

Cytokinesis [,saɪtəʊkɪ'niːsɪs]
The division of the cytoplasm of a cell following the division of the nucleus.

Cytoplasm ['saɪtəʊplæzm]
All cell organelles are located inside the cytoplasm. It is enclosed by the cell membrane. The nucleus is kept separate from the cytoplasm by the nuclear envelope.

Cytoskeleton ['saɪtəʊ,skelɪtn]
Network of fine protein structures that provides the cell with stability and shape. It also functions in intracellular transport.

Dendrogram (evolutionary family tree) ['dendrəgræm]
A diagram that shows the interrelationships between groups of organisms as branching tree-like diagram. The term is mainly used for family trees that are based on genetic features.

Dictyosome ['dɪktəzəʊm]
Cell organelle composed of membrane-bound sacs arranged in stacks. Involved in the export of protein secretions. The entirety of all dictyosomes of a cell is called the *Golgi apparatus*.

Diploid cell ['dɪplɔɪd ,sel]
Cell with two homologous sets of chromosomes (2n) of which one is derived from the mother and the other from the father.

DNA (**d**eoxyribo**n**ucleic **a**cid)
[,diː'en'eɪ] [,diː'ɒksɪ'raɪbəʊnuː,kleɪk 'æsɪd]
Consists of long chains of nucleotides (composed of the sugar deoxyribose, phosphate and one of the 4 bases adenine, thymine, guanine and cytosine). Its structure is often described as a twisted ladder (helix). The genomic information of a cell is encoded within the nucleotide sequence.

DNA probe / hybridisation probe
[,diː'en'eɪ ,prəʊb] [,haɪbrɪdaɪ'zeɪʃn ,prəʊb]
A short, marked, single-stranded DNA (or RNA) molecule (several bases in length) that is used to identify complementary base sequences.

Domestication [də,mestɪ'keɪʃn]
The process by which wild animals and plants become adapted to agriculture/cultivation as a result of the selection by humans.

Dominant ['dɒmɪnənt]
An allele that is expressed even when only one copy is present in an individual, i.e. in a heterozygous condition (see *inheritance*).

Electron microscopy
[ɪ,lektrɒn ,maɪ'krɒskəpɪ]
This microscope uses a beam of electrons instead of a light beam. It allows high magnification and high resolution. This makes the examination of fine structures, viruses and molecules possible.

Electrophoresis [ɪ,lektrəʊfə'riːsɪs]
A technique for separating molecules (proteins, nucleic acids) by their

migration in a solution / gel towards oppositely charged poles in an electric field generated by a direct current. The migration rate varies with the size, shape and charge of the particles.

Endoplasmic reticulum (ER) [ˌendəʊplæzˌmætɪk rɪˈtɪkjələm] [ˌiːˈɑː]
A cell organelle of eukaryotic cells. A complex network of membranous fluid-filled cavities (cisternae). It is involved in the synthesis, conversion and transport of substances.

Endosymbiotic hypothesis [ˌendəʊˌsɪmbaɪˈɒtɪk haɪˌpɒθəsɪs]
A well-established theory concerning the origin of the eukaryotic cell. This hypothesis states that plastids and mitochondria have evolved from prokaryotic (bacteria-like) cells that were taken up but not digested by a nucleated cell and kept as symbionts.

Enzyme [ˈenzaɪm]
Enzymes are proteins that catalyze (i. e. accelerate) biochemical reactions.

Eukaryote [juːˈkærɪəʊt]
Unicellular and multicellular organisms whose cells possess a nucleus and other membrane-bound vesicles, including all members of the protist, fungi, plant, and animal kingdoms.

Eukaryotic cell [juːkærɪˈɒtɪk ˌsel]
A type of cell that is the basic building block of all eukaryotes. It contains, for example, a nucleus and mitochondria, and is highly compartmentalised in contrast to a *prokaryotic cell*.

Feedback [ˈfiːdbæk]
The process in which part of the output of a system is returned to its input in order to regulate its further output (feedback regulation / feedback control loop). Positive feedback tends to reinforce a process, whereas negative feedback causes a slow-down of a process.

Fertilization [ˌfɜːtɪlaɪˈzeɪʃn]
The fusion of the nuclei of two gametes forming a fertilised egg cell (zygote); after *insemination*.

Gamete (sex cell, germ cell) [ˈgæmiːt] [ˈseksˌsel] [ˈdʒɜːmˌsel]
Haploid egg or sperm cell that, in contrast to spores, develops only after *fertilisation* into a multicellular orga-

nism. In *diploid* organisms, gametes are produced in specialised organs (*gonads*, sex glands) by a special form of cell division *(meiosis)*, and in haploid organism by *mitosis*.

Gametophyte [gəˈmiːtəʊfaɪt]
The *haploid* multicellular stage of the plant life cycle producing gametes (germ cells) by mitosis. Fertilised gametes then develop into the spore-producing *diploid* generation *(sporophyte)*. Gametophyte of mosses: gametophore; gametophyte of ferns: prothallium; gametophyte of seed plants: unimpressive group of cells inside the sporophytes.

Gastrula [ˈgæstrʊlə]
A stage in the development of a multicellular organism; it consists of two layers called ectoderm and endoderm enclosing an inner cavity (primitive gut).

Gastrulation [ˌgæstrʊˈleɪʃn]
The stage in embryonic development of a multicellular organism that follows cleavage. The hollow ball *(blastula)* becomes, by invagination or the migration of cells, the *gastrula*.

Gene [dʒiːn]
A hereditary unit consisting of a sequence of DNA that occupies a specific location on a chromosome and codes for a RNA molecule.

Gene expression [ˌdʒiːn ɪkˈspreʃn]
The process by which a gene's coded information is used by a cell; all genes are first transcribed (RNA production) and, if they code for proteins, translation also occurs.

Gene linkage [ˌdʒiːn ˈlɪŋkɪdʒ]
Existence of groups of genes that are inherited together because they are located closely together on the same *chromosome*. *Crossing over* can break up such linkage groups.

Gene mapping [ˌdʒiːn ˈmæpɪŋ]
Determination of the relative positions of different genes on a DNA molecule *(chromosome)*.

Gene pool [ˈdʒiːn ˌpuːl]
The totality of all *genes* in a population at a given time.

Gene regulation [ˌdʒiːn ˌregjəˈleɪʃn]
Regulation of the protein biosynthesis of a cell by gene activation or inhibition.

Gene therapy [ˌdʒiːn ˈθerəpɪ]
Treatment of patients with genetically modified cells: In somatic gene therapy, genes of the affected cells of a patient's body are replaced, supplemented or removed. Germ-line gene therapy of humans, which is the genetic modification of germ cells, is prohibited in Germany for ethical reasons.

Gene transfer [ˌdʒiːn ˈtrænsfɜː]
Methods that are used to insert DNA segments (genes) of any origin into a cell (see *plasmid*).

Genetic code [dʒəˌnetɪk ˈkəʊd]
Encoded instructions for protein biosynthesis; the genetic information of the DNA is translated from a nucleic acid sequence (base sequence) into the amino acid sequence of *proteins*.

Genetic disease (genetic disorder, hereditary disease) [dʒəˌnetɪk dɪˈziːz] [dʒəˌnetɪk dɪˈsɔːdə] [həˌredɪtrɪ dɪˈziːz]
A disease caused by inheritable alterations (mutation) in *genes* or *chromosomes*. A genetic disease can also arise from a spontaneous genetic mutation.

Genetic drift [dʒəˌnetɪk ˈdrɪft]
Random fluctuations in gene or allele frequencies, most evident in small populations (founder effect, bottleneck effect).

Genetic engineering [dʒəˌnetɪk endʒɪˈnɪərɪŋ]
All technologies used to manipulate the genome of a cell by transfer or removal of defined DNA segments for research, medicinal or production purposes.

Genetically induced behaviour / behavioural genetics [dʒəˌnetɪklɪˌɪnˌdjuːst brˈheɪvjə] [bɪˌheɪvjərl dʒəˈnetɪks]
Behavioural traits that do not have to be learnt and instead are based on genetically inherited phylogenetic information.

Genetic fingerprint [dʒəˌnetɪk ˈfɪŋgəprɪnt]
Method for analysing *DNA* from hair, blood stains and other cellular materials in order to identify people. Also called DNA typing and DNA profiling.

Genome [ˈdʒiːnəʊm]
The complete genetic material (coding and non-coding DNA segments) of an organism within one set of chromosomes.

Genomic mutation
[dʒɪˌnəʊmɪk mjuːˈteɪʃn]
Mutation that leads to a change in number of chromosomes within a cell. Single chromosomes can either be gained or lost, or the entire set of chromosomes can be multiplied or halved.

Genotype [ˈdʒenəʊtaɪp]
The entire genetic constitution of an organism.

Germline [ˈdʒɜːmlaɪn]
The sequence of cells in an animal from the fertilised egg cell to the gametes.

Golgi apparatus [ˈgɒldʒɪˌæpəˌreɪtəs]
All *dictyosomes* of a cell are called the Golgi apparatus. It modifies and stores products of the *endoplasmic reticulum*.

Gonads (sex glands)
[ˈgəʊnædz] [ˈseks ˌglændz]
Gamete-producing (sperm, egg cells) organs of animals; female gonads are called ovaries, and male gonads are called testes.

Gonosome (sex chromosomes)
[ˈgɒnəʊsəʊm] [ˈseks ˌkrəʊməsəʊmz]
A chromosome that, in contrast to an *autosome*, carries genes that determine the sex. Gonosomes determine the sex of offspring in the diploid zygote; they are mostly called the X and Y chromosome.

Growth [grəʊθ]
Irreversible increase of body volume by the addition of bodily substance; in multicellular organisms, the growth and proliferation of cells.

Haploid cell [ˈhæplɔɪd ˌsel]
A cell with only one set of chromosomes, for example, a *gamete*.

Hemizygous [ˌhemɪˈzaɪgəs]
In a diploid organism, a hemizygous individual has only one allele for a particular trait; this is especially true for many alleles on gonosomes.

Heterozygous [ˌhetərəʊˈzaɪgəs]
Having two different *alleles* for a specific trait.

HIV (**H**uman **I**mmunodeficiency **V**irus)
[ˌeɪtʃaɪˈviː] [ˈhjuːmən ˌɪmjuːnəʊdɪˈfɪʃənsɪ ˌvaɪərəs]
An RNA virus that causes the immune deficiency disease *AIDS* in humans by infecting immune cells.

Homozygous (true breeding)
[ˌhɒməʊˈzaɪgəs] [ˌtruː ˈbriːdɪŋ]
Having two identical *alleles* for a particular trait (compare *heterozygous*).

Hormone [ˈhɔːməʊn]
A chemical messenger produced in glands and released into the blood to trigger or regulate particular functions of the body.

Hybridisation [ˌhaɪbrɪdaɪˈzeɪʃn]
Molecular biological method; after the heating and subsequent cooling of a mixture of two complementary single strands, the nucleic acid strands of different origin will attach to each other, if they possess sufficient numbers of complementary bases (see also *DNA probe*).

Immunity [ɪˈmjuːnətɪ]
Acquired resistance to a specific disease; the ability of an organism to react against antigens of an infectious agent; adaptive and innate immunity are distinguished.

Incubation period
[ˌɪnkjʊˈbeɪʃn ˌpɪərɪəd]
The time interval that elapses between the introduction of the pathogen and the occurrence of the first clinical signs of disease.

Inheritance [ɪnˈherɪtəns]
The handing down of traits from parents to their offspring. In *recessive-dominant* inheritance, the allele of only one parent determines the phenotype of the heterozygous offspring generation; in *codominant* inheritance, neither phenotype is recessive and hence the heterozygous offspring expresses both phenotypes; in incomplete dominant inheritance, the trait is expressed weakly resulting in an intermediate phenotype in the heterozygous offspring generation.

Insemination [ɪnˌsemɪˈneɪʃn]
Introduction of semen into the genital tract of a female. Insemination occurs prior to *fertilisation*.

Intermediate [ˌɪntəˈmiːdɪət]
Lying between two states; in-between phenotype of heterozygous organisms.

Intron [ˈɪntrɒn]
Non-coding DNA segment within a eukaryotic gene; it is initially transcribed into the respective mRNA segment but is subsequently removed during *splicing* (see *exon*).

Karyogram [ˈkærɪəʊgram]
A diagrammatic representation of the stained metaphase chromosomes of an organism; it enables the size, shape and pattern of the chromosomes to be distinguished.

Ligase [ˈlaɪgeɪz]
Genetic "glue"; enzyme that can link DNA fragment.

Lymph [lɪmf]
The fluid present within the lymphatic system and between tissue cells. It does not contain red blood cells and is a filtrate of the blood in animal species with a closed blood circulation.

Lysis [ˈlaɪsɪs]
The breaking open of a cell or cell organelle by the destruction of its wall or membrane.

Macrophage [ˈmækrəʊfeɪdʒ]
A differentiated cell of the immune system that is specialized in phagocytosis. It engulfs and destroys foreign cells and particles.

Malignant [məˈlɪgnənt]
Cancerous; refers to cells or tumours growing in an uncontrolled fashion.

Marker [ˈmɑːkə]
Substance or structure that is used to detect products, metabolic reactions or genetic modifications.

Matrix [ˈmeɪtrɪks]
Ground substance or ground structure of, for example, chloroplasts.

Meiosis [maɪˈəʊsɪs]
A two-stage type of cell and nuclear division in sexually reproducing organisms. It results in cells with a single set of chromosomes (gametes)

derived from primordial germ cells that have a double set of chromosomes. The genetic information can be recombined during this type of cell division.

Mendelian Laws [menˌdiːlɪən 'lɔːz]
Three laws of inheritance proposed by GREGOR MENDEL (Law of Dominance, Law of Segregation, Law of Independent Assortment).

Meristem ['merɪstem]
An organized undifferentiated plant tissue with rapidly dividing cells that can differentiate to form new tissues or organs.

Metamorphosis [ˌmetə'mɔːfəsɪs]
Hormone-regulated change of shape during the development of an animal. It includes one or several larval stages and optionally a pupa stage.

Metastasis (plural: **metastases**) [mə'tæstəsɪs] [mə'tæstəsiːz]
The spread and growth of cancer (malignant tumour) from its original site to other parts of the body. Usually through the bloodstream or lymphatic system (see *cancer*).

Microtubule ['maɪkrəʊˌtjuːbjuːl]
A tiny hollow tube made of subunits of the tubulin protein. It is important for the formation of the cytoskeleton, the *spindle apparatus* and cilia.

Mitochondrion (plural: **Mitochondria**) [ˌmaɪtəʊ'kɒndrɪən] [ˌmaɪtəʊ'kɒndrɪə]
Eukaryotic cell organelle in which cellular respiration takes place. It is enclosed by a double membrane.

Mitosis [maɪ'təʊsɪs]
Nuclear division during which the two chromatids of a *chromosome* produced during replication are distributed to the newly forming daughter cells. The number of chromosomes per cell remains the same. Nuclear division is usually followed by *cytokinesis*.

Modification [ˌmɒdɪfɪ'keɪʃn]
Change in the appearance (phenotype) of an organism caused by environmental factors. It is not inherited.

Morula (Latin for **"mulberry"**) ['mɔːrʊlə]
An early stage of embryonic development of a multicellular organism. It is characterized by a solid ball of cells that will later develop into the *blastula*.

Mutagens ['mjuːtədʒənz]
Substances that can induce genetic changes in DNA *(mutations)*.

Mutation [mjuː'teɪʃn]
Any qualitative and quantitative change of the genetic material caused by either random spontaneous events or by *mutagens*. Dependent on the extent of the change, *gene*, *chromosome* and *genomic* mutations can be distinguished.

Nucleic acid [nuːˌkleɪɪk 'æsɪd]
A macromolecule composed of nucleotides. It performs several functions in living cells, e.g. the storage of genetic information, its transfer from one generation to the next *(DNA)* and the expression of this information in proteins (via *RNA*).

Nucleolus (plural: **Nucleoli**) [ˌnuːklɪ'əʊləs] [ˌnuːklɪ'əʊlaɪ]
A structure that is found in the *nucleus* and in which ribosomal RNA is transcribed and ribosomal subunits are assembled.

Nucleus (plural: **Nuclei**) ['nuːklɪəs] ['nuːklɪaɪ]
Membrane-bound part of the cell containing the genetic material and controlling growth, metabolism and reproduction. It is only present in eukaryotic cells.

Nucleoside ['nuːklɪəsaɪd]
A precursor of *nucleic acids* consisting of a nitrogen-containing base and a sugar. It is converted into a *nucleotide* by the addition of a phosphate group.

Nucleotide ['nuːklɪətaɪd]
The building blocks of *nucleic acids* (DNA and RNA) and *coenzymes* (ADP, NAD+ and NADP+). Nucleotides are composed of phosphate groups, a sugar molecule and nitrogen-containing bases.

Omnipotency [ˌɒmnɪ'pəʊtənsɪ]
(See *totipotency*).

Oncogene ['ɒnkəʊdʒiːn]
A *gene* that is associated with the development of *cancer*. In cells, it is derived from a "normal" gene that usually controls cell growth (proto-oncogene). It is also present in some *viruses*.

Ontogenesis [ˌɒntəʊ'dʒenəsɪs]
The process of an individual organism growing from the earliest embryonic stage to maturity.

Operon ['ɒpərɒn]
A functionally integrated unit of genes that control gene expression in bacteria. They regulate *enzyme* production.

Organelle [ˌɔːgən'el]
A specialized membrane-bound structure within the cytoplasm that has a defined cellular function.

Parthenogenesis [ˌpɑːθənəʊ'dʒenəsɪs]
A form of reproduction in which egg cells develop into individuals without fertilisation.

PCR (**p**olymerase **c**hain **r**eaction) [ˌpiːsiː'ɑː] [pə'lɪməreɪz 'tʃeɪn rɪˈækʃn]
A method for amplifying a specific DNA segment (basic method in molecular biology).

PGD (**p**reimplantation **g**enetic **d**iagnosis) [ˌpiːdʒiː'diː] [ˌpriːˌɪmplæn'teɪʃn dʒəˌnetɪk ˌdaɪəg'nəʊsɪs]
Screening procedure for multicellular embryos produced by in vitro fertilization (IVF). Cells are removed from the embryo and examined for genetic damage before the embryo is transferred to the mother's womb. PGD is prohibited in Germany because it enables the selection of embryos based on non-medical criteria.

Phage ['feɪdʒ]
Virus that infects microorganisms, e.g. bacteriophage.

Phagocytosis [ˌfægəʊsaɪ'təʊsɪs]
The absorption and digestion of foreign materials by cells (especially cells of the immune system).

Phenotype ['fiːnəʊtaɪp]
The overall attributes of an organism (compare *genotype*).

Plasmid ['plæzmɪd]
A plasmid is a circular loop of DNA found in prokaryotic cells (e.g. bacteria). It is independent of the rest of the cellular genome. Bacteria can transfer plasmids to other bacteria and they can be used for gene transfer (see *vector*).

Plastid [ˈplæstɪd]
A double-membrane-bound cyto-plasmic organelle containing its own DNA and 70 S ribosomes. It is found in plants and algae. Chloroplasts are the main type of plastids, but chromo-plasts and leucoplasts also occur.

Point mutation [ˌpɔɪnt mjuːˈteɪʃn]
The change or loss of a single base in a DNA molecule (see *mutation*).

Polygenic inheritance
[ˌpɒlɪˌdʒenɪk ɪnˈherɪtəns]
A trait or characteristic that is influ-enced by the expression of more than one *gene*.

Polymorphism [ˌpɒlɪˈmɔːfɪzm]
The existence of two or more alterna-tive forms (alleles) of a gene result-ing in the occurrence of many forms within the same species, e.g. shape or colour variations.

Polyploid [ˈpɒlɪplɔɪd]
Organism with more than two full sets of homologous chromosomes in its cells.

Precipitation [prɪˌsɪpɪˈteɪʃn]
The clumping of molecules by bridging molecules (e.g. antibodies).

Primer [ˈpraɪmə]
A short single-stranded *nucleotide* chain that has complementary bases to the template (e. g. DNA). It is used to initiate the synthesis of the comple-mentary strand by polymerases (e. g. in *PCR*, *replication*).

Prion [ˈpraɪɒn]
An infectious protein agent responsi-ble for degenerative diseases of brain tissue (e.g. BSE).

Prokaryote [prəʊˈkærɪəʊt]
A single-celled organism lacking a true nucleus, such as eubacteria and archaebacteria (compare *eukaryote*).

Prokaryotic cell [ˌprəʊkærɪˌɒtɪk ˈsel]
Cell type found in all *prokaryotes*. It lacks a nucleus (compare *eukaryotic cell*).

Proliferation [prəˌlɪfəˈreɪʃn]
A characteristic of life. The number of individuals is increased. Proliferation is always connected to *reproduction*.

Promotor [prəˈməʊtə]
A regulatory region of DNA that is located at the start of a gene and that functions as the starting sequence for the DNA polymerase that produces mRNA (see *transcription*).

Protein [ˈprəʊtiːn]
A large molecule composed of a long chain of amino acids arranged in a 3-dimesional structure. Proteins perform a wide variety of functions as enzymes, structural components or contractile filaments.

Protein biosynthesis
[ˌprəʊtiːn ˌbaɪəʊˈsɪnθəsɪs]
The process of making *proteins* in cells according to the information con-tained on the *DNA*. It is divided into two main stages: *transcription* and *translation*.

Protists [ˈprəʊtɪsts]
Eukaryotic unicellular, colonial or multicellular organisms that do not belong to fungi, plants or animals. A kingdom of organisms additional to prokaryotes, animals, plants and fungi. Examples: diatoms, green algae and amoeba.

Reaction norm [rɪˈækʃnˌnɔːm]
In genetics: range of phenotypes that are derived from the same genotype. Organisms can vary in appearance because of environmental influences, even though they have the same genetic set-up.
In behavioural science: range of the reactions of an animal to directing stimuli.

Receptor [rɪˈseptə]
A molecule that occurs on the surface of a cell or within a cell and that re-cognises other specific molecules to initiate cellular processes.

Recessive [rɪˈsesɪv]
An allele that is not expressed in *hete-rozygous* organisms (see *inheritance*).

Recombination [ˌriːkɒmbɪˈneɪʃn]
The exchange (reshuffling) and new combining of genetic material, e.g. during meiosis. Recombination leads to genetic variants within the same species and is an important evolutio-nary factor.

Replication [ˌreplɪˈkeɪʃn]
Identical (semi-conservative) duplica-tion of the DNA prior to every cell divi-sion (*mitosis* and *meiosis*). The DNA double strand is separated and the parental single strands are each used as the template for a daughter strand.

Reproduction [ˌriːprəˈdʌkʃn]
A fundamental characteristic of life. Independent individuals of the same species are produced by the act of passing on genetic information. Sexual reproduction includes the fertilisation of gametes. Asexual reproduction is a process in which new organisms develop from a part of the parent organism. Reproduction is mostly linked with multiplication / an increase in number.

Resistance [rɪˈzɪstəns]
The inherited ability to withstand diseases, climate, toxins, drugs, and much more. Resistance should be di-stinguished from acquired resistance (see *immunity*).

Restriction enzyme
[rɪˈstrɪkʃn ˌenzaɪm]
An enzyme that breaks double-stranded DNA at highly specific base sequences. Basic tool of *genetic engi-neering* (compare *ligase*).

Retrovirus [ˈretrəʊˌvaɪərəs]
A type of virus that contains single-stranded RNA as its genetic material (e.g. *HIV*). Prior to replication in a host cell, the RNA is translated into DNA by the enzyme reverse transcriptase.

Reverse transcriptase
[rɪˌvɜːs trænˈskrɪpteɪz]
Enzyme that synthesizes DNA directly from RNA.

R$_f$-value (**R**etention factor)
[ˌɑːˈref ˌvæljuː] [rɪˈtenʃn ˌfæktə]
A chromatographic value that is the quotient of the distance moved (run length) by the compound divided by the distance moved by the eluent (front).

Rhesus factor [ˈriːsəs ˌfæktə]
An antigen (also referred to as antigen D or Rh) on the surface of red blood cells. In contrast to the ABO system, D-antibodies develop in Rh-negative people only after their having contact with Rh-positive blood.

Ribosome [ˈraɪbəʊsəʊm]
Cellular organelle that is composed of proteins and *RNA* and that is the site of protein synthesis (translation).

RNA (Ribo**n**ucleic **a**cid)
[ˌɑːrenˈeɪ] [ˌraɪbəʊnuːˌkleɪɪk ˈæsɪd]
In contrast to DNA, a short single-stranded chain of *nucleotides* that contains the sugar ribose, a phosphate group and the bases adenine, uracil, guanine and cytosine. Three types of RNA are distinguished:
1) messenger-RNA (mRNA) is used for transcription,
2) transfer-RNA (tRNA) is used for translation and
3) ribosomal-RNA (rRNA) forms a part of ribosomes.
The genome of retroviruses is composed of RNA.

Sexual dimorphism
[ˌsekʃʊəl daɪˈmɔːfɪzm]
The existence of differences regarding size, shape, colour, physiology or behaviour (secondary sexual characteristics) between sexes of the same species.

Splicing [ˈsplaɪsɪŋ]
The removal of segments from pre-mRNA that then leaves as the nucleus of eukaryotic cells as mature mRNA. The removed segments represent the *introns*.

Spore [spɔː]
A reproductive cell that develops into a multicellular individual without fertilisation. Spores are produced in specialised tissue by mitosis (mitospores) or meiosis (meiospores). They can be either haploid or diploid. Bacterial resting structures are also called spores (endospores).

Sporophyte [ˈspɔːrəʊfaɪt]
A multicellular diploid generation in the life cycle of a plant. It produces (meiotic) spores. The unfertilised spores then develop into a gamete-producing haploid generation *(gametophyte)*. Sporophyte of mosses: long-stalked spore capsule, sporophyte of ferns: fern plant, sporophyte of seed plants: visible plant.

Stem cell [ˈstem ˌsel]
An undifferentiated cell that is found in the tissue of multicellular organisms and that is able to divide thereby making growth and the renewal of tissues possible.

Symbiosis [ˌsɪmbaɪˈəʊsɪs]
The close association of two or more different organisms of different species in which both benefit from the relationship.

Syndrome [ˈsɪndrəʊm]
A group of symptoms that occur together and represent a particular disease.

Totipotency (omnipotency)
[ˌtəʊtɪˈpəʊtənsɪ]
Ability of specific cells *(stem cells)* to produce an entire organism/any cell type of an entire organism by cell division and differentiation.

Transcription [trænˈskrɪpʃn]
The transfer of genetic information from DNA into mRNA. Part of protein synthesis.

Transduction [trænzˈdʌkʃn]
In genetics: The transfer of foreign DNA by *phages* to a *bacterium*.
In physiology of the sensory system: the initiation of electrical excitation caused by stimuli in the sensory cells.

Transformation [ˌtrænsfəˈmeɪʃn]
Transfer of *DNA* to any living cell.

Transgenic organism
[trænzˌdʒenɪk ˈɔːgənɪzm]
Organism that has received genes foreign to its species by methods of genetic engineering.

Translation [trænzˈleɪʃn]
Polypeptide synthesis in ribosomes based on an mRNA strand. Part of protein synthesis.

Translocation [ˌtrænzləʊˈkeɪʃn]
A piece of one chromosome breaks off and becomes attached to another chromosome.

Transposition [ˌtrænspəˈzɪʃn]
The movement of DNA segments within the genome (jumping genes).

Triplet [ˈtrɪplɪt]
A sequence of three nucleotides comprising an information unit of RNA or DNA (see *codon*).

Trisomy [traɪˈsəʊmɪ]
Genomic mutation leading to the presence of three instead of two chromosomes in cells of the body. This leads to phenotypic changes, e. g. Turner syndrome, Klinefelter syndrome.

Tumour (swelling) [ˈtjuːmə]
Undifferentiated cell mass that can form because of the unregulated growth of one cell in the tissue. Benign tumours grow slowly; malignant tumours (cancer) can spread throughout the body.

Variability [ˌveərɪəˈbɪlətɪ]
Difference between individuals within a population. It can be caused genetically (genetic variation, *polymorphism*) or by environmental factors (environmental variation).

Vegetative [ˈvedʒɪtətɪv]
Means "not reproducing sexually", asexual reproduction.

Vector [ˈvektə]
A DNA molecule functioning as "gene vehicle" in which foreign DNA can be integrated (see *restriction enzyme*, *ligase*) and then replicated in the host cell (see *plasmid*).

Virus [ˈvaɪərəs]
Non-cellular genetic unit made up of *nucleic acids* and *proteins* that can only reproduce within a host cell.

Zygote [ˈzaɪgəʊt]
A diploid cell formed by the union of a male sex cell (a sperm) and a female sex cell (an ovum). The zygote develops into the embryo following the instructions encoded in its genetic material, the DNA. First cell of a new organism.

Pictures sources

Cover FOCUS (EOS/Meckes), Hamburg – 6.1 Avenue Images, Hamburg – 6.2 Mauritius (Botanica/Alexandra Grablewski), Mittenwald – 7.1 Helga Lade (BAV), Frankfurt – 7.2 Okapia (NAS, W.&D.McIntyre), Frankfurt – 7.3 FWU, Grünwald – 7.4 FOCUS (SPL), Hamburg – 8.Rd. The Rockefeller Center Archive, New York – 9.S MEV, Augsburg – 13.Rd. aus „Benjamin Lewin, Genes IV" (S. 329), Oxford University Press (Bernard Hirt) – 14.S FOCUS (Dr. Gopal Murti/Science Photo Library), Hamburg – 14.3 aus „Henning, Genetik (S. 67), Springer-Verlag, Heidelberg – 17.I1 Okapia (Norbert Lange), Frankfurt – 17.I2 FOCUS (Prof. P. Motta & T. Naguro/SPL), Hamburg – 19.S FWU, Grünwald – 22.2 FOCUS (SPL), Hamburg – 26.Rd. DSMZ Deutsche Sammlung von Mikroorganismen und Zellkulturen, Braunschweig – 31.S MEV, Augsburg – 32.Rd. Okapia (Hans Reinhard), Frankfurt – 33.S MEV, Augsburg – 33.1 Reinhard-Tierfoto, Heiligkreuzsteinach – 33.2 Angermayer (Tierpark Hellabrunn), Holzkirchen – 33.5 Okapia (Carl Roessler), Frankfurt – 34.1 Okapia (Klaus von Mandelsloh), Frankfurt – 34.Rd. Dr. Donald E. Keith; Stephenville – 36.Rd. Albert Bonniers Förlag AB (Lennart Nilsson), Stockholm – 38.Rd.1 Deutsches Museum, München – 38.Rd.2 Klett Archiv (Dr. Horst Schneeweiß), Stuttgart – 40.2 Max-Planck-Institut für Züchtungsforschung (Maret-Linda Kalda), Köln – 41.S Corel Corporation, Unterschleissheim – 43.1a K. Hägele – 44.S FOCUS (Jürgen Berger), Hamburg – 44.1, 2, 3a/b FWU, Grünwald – 46.1 aus „E. Passarge, Taschenatlas der Genetik", S. 81, Georg Thieme Verlag, Stuttgart – 49.S FOCUS (Pascal Goetgheluck), Hamburg – 51.S Angermayer (Hans Pfletschinger), Holzkirchen – 53.1 aus „Alberts/Bray/Lewis/Roberts/Watson, Molekularbiologie der Zelle", S. 1287, VCH-Verlag, Weinheim – 55.S Albert Bonniers Förlag AB (Lennart Nilsson), Stockholm – 55.2a/b aus „K. V. Hinrichsen (Hrsg.): Humanembryologie", Springer Verlag Berlin, Heidelberg 1990 – 58.S FOCUS (Jürgen Berger), Hamburg – 58.1, 4, 5 Deutsches Museum, München – 58.2, 3 Corbis (Bettmann), Düsseldorf – 59.1 Corbis (Bettmann), Düsseldorf – 59.2 aus „Medical News", 1/62 – 59.3 Deutsches Museum, München – 59.4 Picture-Alliance (dpa/Zentralbild, Klaus Franke), Frankfurt – 61.S dpa, Frankfurt – 64.S (Heiner Heine) – 65.S Manfred P. Kage / Christina Kage, Lauterstein – 65.2 Helga Lade (NDS), Frankfurt – 66.1 FOCUS (Lauren Shear/SPL), Hamburg – 66.2 FOCUS (Science Photo Library/L. Willatt), Hamburg – 68.S Okapia (Lond. Sc. Films, OSF), Frankfurt – 68.1 Imago Stock & People (Horst Rudel), Berlin – 70.Rd.1, 3 Albert Bonniers Förlag AB (Lennart Nilsson), Stockholm – 70.Rd.2 FOCUS (Science Pictures, Science Photo Library), Hamburg – 71.S Albert Bonniers Förlag AB (Lennart Nilsson), Stockholm – 72.Rd. Klett Archiv (Dr. Horst Schneeweiß), Stuttgart – 76.2a MEV, Augsburg – 76.2b Klett Archiv (Dr. Goran Söhl), Stuttgart – 77.K FOCUS (Dept. of Clinical Cytogenetics), Hamburg – 80.Rd. Okapia (NAS/H. Morgan), Frankfurt – 82.S Tilman Wischuf, Cleebronn – 82.1 Mauritius (Rossenbach), Mittenwald – 82.2, 3 Prof. Jürgen Wirth, Dreieich – 83.3 Hansjörg Neth, Troy, USA – 84.1 FOCUS (EOS/Meckes), Hamburg – 89.K The Nobel Foundation, Stockholm – 90.Rd. Okapia (D. Scharf, P. Arnold), Frankfurt – 92.S Albert Bonniers Förlag AB (Lennart Nilsson), Stockholm – 93.S FOCUS (National Cancer Institute / SPL), Hamburg – 96.Rd.1, Rd.2 Catherine Botteron, Bern – 97.S FOCUS (EOS), Hamburg – 98.S Okapia (T. McHugh), Frankfurt – 98.1 AKG, Berlin – 99.1 Mauritius (Torino), Mittenwald – 100.2 Okapia (Ulrich Zillmann), Frankfurt – 101.1 Corbis (Roger Tidman), Düsseldorf – 101.2 Topic Media (Volkmar Brockhaus), Ottobrunn – 101.3 Topic Media (Fleetham), Ottobrunn – 102.1a, 2 Tierbildarchiv Angermayer (H. Pfletschinger), Holzkirchen – 102.1b aus „Heribert Schmid, Wie Tiere sich verständigen", Otto Maier Verlag, Ravensburg, S. 24 (M. Boppré) – 103.1 Corbis (Barton), Düsseldorf